TURING

程序员的英语

[韩] 朴栽浒　李海永　著　[美] Kevin Steely　审校

颜廷连　译

人 民 邮 电 出 版 社

北 京

图书在版编目(CIP)数据

程序员的英语 /(韩)朴栽浒,(韩)李海永著;颜廷连译. -- 北京:人民邮电出版社,2018.3
ISBN 978-7-115-47305-9

Ⅰ. ①程… Ⅱ. ①朴… ②李… ③颜… Ⅲ. ①程序设计一英语 Ⅳ. ①TP311.1

中国版本图书馆CIP数据核字(2017)第284374号

内 容 提 要

本书旨在最大限度提高对开发人员最重要的英文读解能力，培养能够快速阅读英文报道等IT相关文档的基本技能。书中借助最新IT媒体风格的报道，提供有趣且有益的知识，详细解释英语技术术语。课后附有英文思维训练，内容和形式都贴合一线程序员需求，能够激发阅读兴趣。例文难易得当，结构安排合理，是技术学习与英语学习的有机结合。

◆ 著　　　　[韩] 朴栽浒　李海永
审　　校　[美] Kevin Steely
译　　　　颜廷连
责任编辑　陈　曦
责任印制　周昇亮

◆ 人民邮电出版社出版发行　北京市丰台区成寿寺路11号
邮编　100164　电子邮件　315@ptpress.com.cn
网址　https://www.ptpress.com.cn
廊坊市印艺阁数字科技有限公司印刷

◆ 开本:880×1230　1/32
印张:9　　2018年3月第1版
字数:343千字　　2025年1月河北第26次印刷

著作权合同登记号　图字:01-2016-1579号

定价:49.00元

读者服务热线:(010)84084456-6009　印装质量热线:(010)81055316

反盗版热线:(010)81055315

广告经营许可证:京东市监广登字20170147号

版权声明

前　言

高中时，我们曾比任何人都努力学习英语，大学时也读过很多原著。但上班后，蓦然间你会发现，不知何时已经失去了对英语的感觉。这其中必然有很多原因，但我想最大的问题应该还是缺乏练习。英语作为一门外语在日常生活中并不常用，如果缺乏有意识的学习，则很容易忘记语法和单词。但重拾高中语法和阅读理解已非易事，因为教材和现实毕竟有很大差距。不过，只阅读与自身专业相关的英文指南并不能提高英语水平，应该借助相关领域的专业知识举一反三。

究竟该如何做呢？我们迫切期待可以看懂国外各大 IT 新闻网站及知名博主的帖子，而现实往往不尽如人意；我们知道自己迫切需要达到这个水平，却苦于找不到方法。关于如何提高应试成绩或会话能力的建议铺天盖地，但很少有能紧跟 IT 趋势并提高阅读能力的手段。在现代社会生活异常繁忙的当下，难道没有高效的方法吗？

本书旨在解决读者的上述苦恼。我们在教书育人的过程中，翻译过诸多 IT 专业书籍（包括《黑客》韩文版在内的高难度图书），也曾苦恼该如何才能提高 IT 人才的英语阅读能力。最终，在 Hanbit Media 专业图书组的积极协助下，本书得以问世。这本书既是工具书（收录语法和单词 / 例句），又极具实用功能（例文中的句子是实际 IT 业务中会出现的各种信息），读者群面向关注 IT 并希望搜集英语 IT 信息的高中生、大学生、准备就业的学生、职场人士。

书中例文选取 IT 领域中人气攀升的安全 / 黑客攻击、无人机 / 机器人、大数据、物联网、云为主题，难易适中，生动有趣，最大限度吸引读者阅读。文后全面整理了相关语法、单词、例句等，帮助提高学习效率。各单元均配有习题（造句、阅读理解、讨论），读者可自测。本书按星级分为不同难易度，以满足不同层次读者的需求。书中收录国外 IT 新闻网站和知名博客中常见的实用单词和例句，相信通过日积月累的练习，读者能逐渐阅读相关 IT 新闻。各位可通过最后一部分难度最高的文章，对自己的英语水平提升状况进行全面把握。

对于初级读者，我们更建议由易渐难阅读本书，而不是根据单元设置顺序进行学习（请参考书后难易度目录）。MP3 下载地址为“图灵社区”本书主页（http://www.ituring.com.cn/book/1770）“随书下载”部分。

在此，对孜孜不倦校对全书例文、语法、例句、单词的 Kevin 以及何俊表示感谢。

朴栽浒・李海永

2015 年 8 月

本书用法

各位可以采用以下两种方式阅读本书：很久没有接触过英语的读者可参照 A 方式，对英语语法留有印象或者正在同时学习英语语法的读者可参照 B 方式。

A 方式：以语法为中心

Step 1

首先学习“核心语法”，把握例文中出现率较高的语法。

Step 2

借助“核心语法”迅速浏览全文，暂时略过生词，整体把握全文即可，在 Step 3 集中攻破生词。

B 方式：以阅读理解为中心

Step 1

不要在乎语法及单词，首先快速通读全文。

Step 2

对照“核心语法”进行学习。例文中有些语法会重复出现，如遇到漏学的部分，可重回 Step 1 学习。Step 3 集中攻破生词。

Step 3

可通过“单词 & 短语”和“技术术语”查看单词。本书分别列出熟词和生词，并为生词附加了练习。同时，利用常用单词造句，反复练习并背诵将有助于学习。

Step 4

可根据需要解答“根据提示完成句子”和“思考题”。如果常用英语写信，建议多关注“根据提示完成句子”。

Step 5

如果参加了学习小组等集体学习活动，建议挑战“讨论”题。

Step 6

参照“翻译”进行自测。

各单元结构

本书选取人气攀升的安全 / 黑客攻击、无人机 / 机器人、大数据、物联网、云等 IT 领域为主题，例文难易适中，生动有趣。书中选取难易适中的单词和语法以提高阅读速度，单词难度从一星到四星逐步加大；例句涵盖单句和复句，形式多样。

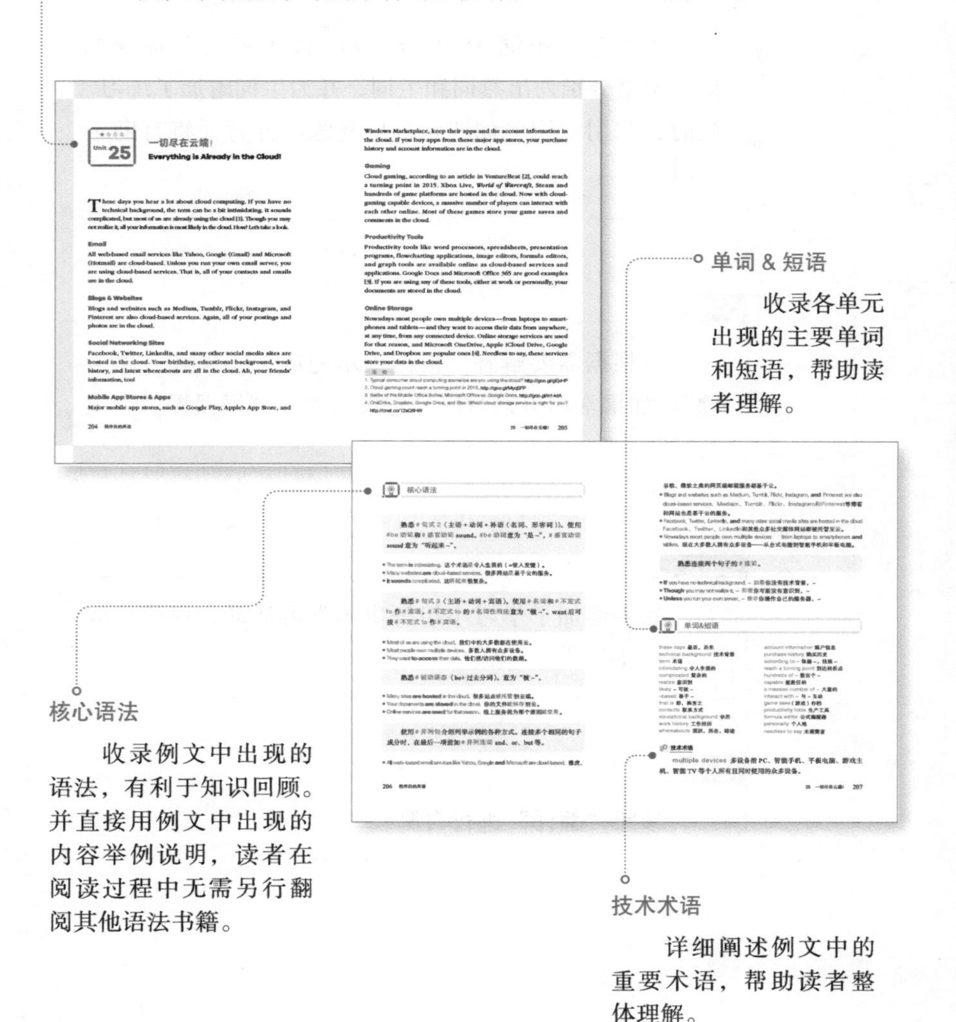

Unit 25

一切尽在云端！

Everything is Already in the Cloud!

These days you hear a lot about cloud computing. If you have no technical background, the term can be a bit intimidating. It sounds complicated, but most of us are already using the cloud [1]. Though you may not realize it, all your information is most likely in the cloud. How? Let's take a look.

Email

All web-based email services like Yahoo, Google (Gmail) and Microsoft (Hotmail) are cloud-based. Unless you run your own email server, you are using cloud-based services. That is, all of your contacts and emails are in the cloud.

Blogs & Websites

Blogs and websites such as Medium, Tumblr, Flickr, Instagram, and Pinterest are also cloud-based services. Again, all of your postings and photos are in the cloud.

Social Networking Sites

Facebook, Twitter, LinkedIn, and many other social media sites are hosted in the cloud. Your birthday, educational background, work history, and latest whereabouts are all in the cloud. Ah, your friends' information, too!

Mobile App Stores & Apps

Major mobile app stores, such as Google Play, Apple's App Store, and Windows Marketplace, keep their apps and the account information in the cloud. If you buy apps from these major app stores, your purchase history and account information are in the cloud.

Gaming

Cloud gaming, according to an article in VentureBeat [2], could reach a turning point in 2015. Xbox Live, *World of Warcraft*, Steam and hundreds of game platforms are hosted in the cloud. Now with cloud-gaming capable devices, a massive number of players can interact with each other online. Most of these games store your game saves and comments in the cloud.

Productivity Tools

Productivity tools like word processors, spreadsheets, presentation programs, flowcharting applications, image editors, formula editors, and graph tools are available online as cloud-based services and applications. Google Docs and Microsoft Office 365 are good examples [3]. If you are using any of these tools, either at work or personally, your documents are stored in the cloud.

Online Storage

Nowadays most people own multiple devices—from laptops to smartphones and tablets—and they want to access their data from anywhere, at any time, from any connected device. Online storage services are used for that reason, and Microsoft OneDrive, Apple iCloud Drive, Google Drive, and Dropbox are popular ones [4]. Needless to say, these services store your data in the cloud.

204 程序员的英语 | 25 一切尽在云端！ 205

核心语法

单词&短语

技术术语

multiple devices 多设备指 PC、智能手机、平板电脑、游戏主机、智能 TV 等个人所有且同时使用的众多设备。

206 程序员的英语 | 25 一切尽在云端！ 207

单词 & 短语

收录各单元出现的主要单词和短语，帮助读者理解。

核心语法

收录例文中出现的语法，有利于知识回顾。并直接用例文中出现的内容举例说明，读者在阅读过程中无需另行翻阅其他语法书籍。

技术术语

详细阐述例文中的重要术语，帮助读者整体理解。

根据提示完成句子

利用例文内容设计填空题，让读者进行写作练习。“好记性不如烂笔头”。

思考题

各单元配有简单的问题和有价值的相关技术问题，帮助读者自测理解情况，题后配有参考答案。

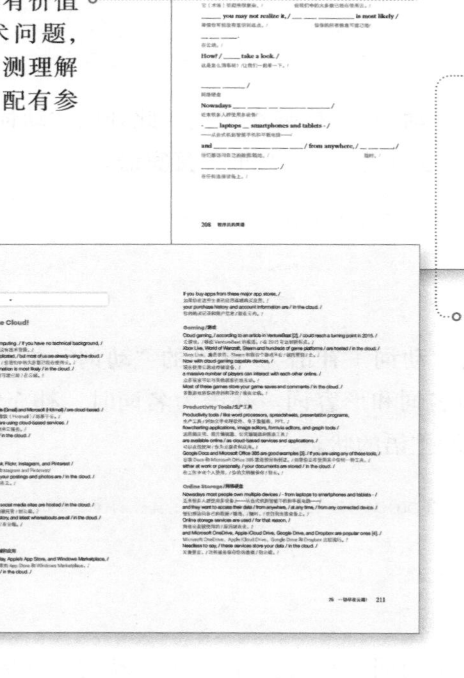
根据提示完成句子

思考题

翻译

Everything is Already in the Cloud!

一切早在云端!

讨论

根据例文内容，各单元选取相关话题供读者讨论。“讨论”题难度比“解答题”更高，是值得期待的挑战。

翻译

课后配有翻译，以检验读者对各单元内容的理解状况。本书采用原文和译文一一对应的方式，帮助各位更准确把握每个单词的含义。

五种句式

英语的 5 种句式曾在日本大受追捧，后来被引入韩国，逐渐成为基本语法句式。当然，理解关系复杂的复句时，仅靠这 5 种语法略显不足，但通过断句方法学习还是有价值可言的。

有些畅销字典将句型细分为 25 种，借助这些细分句型学习既增加负担又很难理解。因此，本书将围绕这 5 种基本的语法句式进行说明。

句式1

句式 1 为〈主语 + 动词〉，是基本句式。此处的“动词”指完全不及物动词（字典中常标注为 vi.），后面不接宾语。

He made toward the central station. 他向中央站走去。

句式2

句式 2 为〈主语 + 动词 + 补语〉。此处的“动词”指不完全不及物动词，补语可以为名词和形容词。补语为名词时，补充说明主语；补语为形容词时，表示主语的状态。

He will make a good programmer. 他将成为一名优秀的程序员。

句式3

句式 3 为〈主语 + 动词 + 宾语〉。此处的“动词”指完全及物动词（字典内常标注为 vt.）。句式 2 和句式 3 最简单的区分方式是：如果一般动词可以用 be 动词代替，则属于句式 2，反之属于句式 3。

Make hay while the sun shines. 晒草要趁太阳好。（比喻抓紧时机行事）

句式4

句式 4 为〈主语 + 动词 + 宾语 + 宾语〉。此处有两个宾语，前一个宾语为间接宾语，后一个宾语为直接宾语。该句式表给予时使用。

He made me the simple sudoku app. 他给我做了个简单的数独APP。

句式5

句式 5 为〈主语 + 动词 + 宾语 + 补语〉。需要注意，该句式中同时存在宾语和补语。换言之，宾语和补充说明宾语的补语同时出现。名词作宾补时，宾语和宾补指同一人或物；形容词作宾补时，指宾语的形态或性质。宾补可以是不定式 to、现在分词、过去分词。

He made her happy. 他使她幸福。

如上所示，我们使用 make 例句表现了 5 种句式。一般动词不适用于所有句式，所以各位在学习过程中要多关注动词用法。

目 录

安全/黑客攻击

无人机/机器人

大数据

物联网

云

实战

第一部分

安全/黑客攻击

安全和黑客攻击一直以来都是新闻报道和报刊杂志老生常谈的话题，它们不仅会招致恐慌，更会带来实质性的危害。本部分选取常见的术语和概念，为大家讲述安全和黑客攻击的有趣故事。

谷歌黑客精英

Google's Elite Security Team, Project Zero

When Apple launched the iPhone through an exclusive partnership with AT&T in 2007, seventeen-year-old George Hotz wanted to use an iPhone but not with AT&T. He wanted to make calls through his own T-Mobile network, so he cracked AT&T's lock on the iPhone [1]. Apple scrambled to fix the bug that allowed this, but officially ignored George Hotz.

Later in 2011, Hotz reverse engineered the Playstation 3 and posted a copy of the root keys on his website. Sony sued him but settled after Hotz promised never to hack Sony products again.

Then in early 2014, at Google's hacking competition, Hotz discovered a security hole in Google's Chrome OS. The company gave him a $150,000 reward. Two months later, Google's security engineer Chris Evans offered him a position in a team of elite hackers. George Hotz accepted the offer and now works for Google's security team *Project Zero* [2].

Project Zero worked in secret until Google publically revealed the team in July 2014. Its sole mission is tracking down and getting rid of security flaws in the world's software. These flaws are called *zero-day* vulnerabilities, which are a common target of cyber criminals.

Project Zero's hackers aren't just looking into the products that Google makes. They are free to hack any software in the world. Why? They want to make a safer Internet for everyone. The team's policy is simple. The team notifies vendors of vulnerabilities immediately. If fixes are not available within 90 days, bug reports automatically become available to the public. The 90-day disclosure policy appears to be working in most cases. The Adobe Flash team fixed 37 Project Zero vulnerabilities (or 100%) within the 90-day period. The Project Zero blog indicates that 85% of all vulnerabilities are patched before the deadline [3].

However, recently Google's strict 90-day policy came under fire from Microsoft and Apple. The Project Zero team publicly disclosed bugs which were present in Windows 8.1 and MacOS X before Microsoft and Apple released patches. Microsoft heavily criticized Google since the company was scheduled to release a patch just two days later [4]. Recently Google loosened its 90-day policy with an additional 14-day grace period. Now vendors have an additional 14 days to patch vulnerabilities as long as they inform Google of the release schedule before the deadline.

"People deserve to use the Internet without fear that vulnerabilities out there can ruin their privacy with a single website visit. We're going to try to focus on the supply of these high value vulnerabilities and eliminate them." says Evans.

出 处

1. Geroge Hotz, Wikipedia, http://goo.gl/V6bl
2. Meet 'Project Zero,' Google's Secret Team of Bug-Hunting Hackers, http://goo.gl/E8rEHy
3. Project Zero, http://goo.gl/dx46YY
4. Windows: Elevation of Privilege in ahcache.sys/NtApphelpCacheControl, Google Security Research, https://goo.gl/c2qgFx

核心语法

熟悉 # 句式 4〈主语 + 动词 + 间接宾语 + 直接宾语〉，# 句式 4 的动词可以理解为“给某人～、使某人做某事、让某人～”。鉴于 # 句式 3 和 # 句式 4 的动词较易混淆，此处将同时介绍句式 3 中的动词。

- The company **gave** him a reward. 公司给他奖励。
- Chris Evans **offered** him a position. 克里斯・埃文斯提供给他一个职位。
- The team **notifies** vendors **of** vulnerabilities. 团队向厂商通报漏洞。
  The team notifies vendors vulnerabilities. (×)
- They **inform** Google **of** the release schedule. 他们向谷歌告知上线日程。
  They inform Google the release schedule. (×)

熟悉 # 关系代名词 that 和 which。# 关系代名词连接两个句子，兼具 # 连词和 # 代名词的作用。根据 # 先行词种类和 # 格的不同，# 关系代名词也相应发生变化。# 先行词是修饰 # 关系代名词从句的名词。# 格在 # 关系代名词从句中发挥作用。# 关系代名词有一定规律可循，理解原理并参照例句学习将事半功倍。

- The team disclosed **bugs**. + **They** were present in Windows 8.1. 团队公布了Bug。+ 它们存在于Windows 8.1中。
  = The team disclosed **bugs** (**that** were present in Windows 8.1). 团队公布了Bug/存在于Windows 8.1中的。
  = The team disclosed **bugs** (**which** were present in Windows 8.1).
  ▶先行词bugs指物，主语是they，此时用主格关系代名词that或which。
- The hackers are looking into **the products**. + Google makes **them**. 黑客们调查产品。+ 谷歌制造了它们。
  = The hackers are looking into **the products** (**that** Google makes). 黑客们调查产品/谷歌制造的。

= The hackers are looking into **the products** (**which** Google makes).

▶先行词the products指物，宾语为them，此时用宾格关系代名词that或 which。

熟悉 # 关系代名词引导非限定性定语从句。这类词用于补充说明前文内容。# 关系代名词前加（,）可理解为〈连接词 + 代名词〉。

- These flaws are called ***zero-day* vulnerabilities**, **and they** are a common target of cyber criminals. 这些漏洞被称为“零日漏洞” + 它们是网络罪犯常攻击的目标。

= These flaws are called ***zero-day* vulnerabilities**, **which** are a common target of cyber criminals. 这些漏洞被称为“零日漏洞”，/它们是网络罪犯常攻击的目标。

▶先行词vulnerabilities指物，主语为they，此时用主格关系代名词which。引导非限定性定语从句时不能用that。

单词&短语

launch 发行
exclusive partnership 独家合作伙伴
crack a code 破译密码
scramble to V 争先恐后做某事
fix a bug 修复 Bug
officially 官方地
ignore 忽视
reverse-engineer 逆向工程
sue 控诉
settle 达成协议，定居
hacking competition 黑客攻击大赛
security hole 安全漏洞
security engineer 安全工程师
accept an offer 接受提案
in secret 秘密地
publically 公开地
reveal 揭露
sole 唯一的
track down 追踪
get rid of 消除、摆脱
security flaw 安全漏洞
look into 调查 ~
be free to V 自由地做某事
policy 政策
notify A of B 向 A 告知 B
automatically 自动地
disclosure policy 公布政策
appear to V 看起来像 ~
deadline 截止日期
patch 补丁，修补

strict policy 严苛的政策
come under fire 遭到攻击，受到谴责
present 目前的、现在的
release 发布、发行
as long as 只要 ~
inform A of B 向 A 通告 B
deserve to V 值得做 ~，有资格做 ~
ruin 摧毁、破坏
focus on 专注于
supply 提供

根据提示完成句子

Project Zero worked / __ ______ /
Project Zero运作 / 秘密地/

until Google __________ ________ the team / in July 2014. /
直到谷歌对外公布这个团队 / 在2014年 7月。

Its ____ _______ is ________ ____ /
它唯一的工作就是追踪/

and _______ ___ __ _________ _____ / in the world's software. /
和解除安全漏洞/在全球软件中。

These _____ are called ____ -___ ___________, /
这些漏洞被称为“零日漏洞”，

which are a ______ ______ / of _____ _________. /
它们是常攻击的目标 /网络罪犯。

Project Zero's hackers aren't just _______ ____ the products /
Project Zero的黑客并不只调查产品/

that Google makes. /
谷歌制造的。/

They are ____ __ ____ any software / in the world. / Why? /
他们自由入侵任何软件/世界上的。 /为什么？ /

They want to make a safer Internet /
他们想构建一个更安全的网络/

for everyone. / The team's ______ is simple. /
为所有人。 /这个团队的策略很简单。

The team notifies vendors / of vulnerabilities / immediately. /
他们通知厂商 /漏洞 /立即。

思考题

解答题

1 （理解）以下内容中，哪一项与Project Zero无关？

ⓐ 减少“零日漏洞”危害的项目

ⓑ 在政府援助下进行

ⓒ 若 90 天内不修改 Bug 将自动公之于众

ⓓ 谷歌构建并运营

2 （论述）谷歌入侵自家公司软件及其他公司软件的原因何在？

讨论

1 若开发软件的团队或者个人不配合，Project Zero 可能转变为具有攻击性的“零日漏洞”，使软件陷入危险。针对这种可能性展开讨论。

课堂总结

1 学习易与#句式4动词混淆的#句式3动词。

2 学习#关系代名词的概念和种类。

3 学习#限定性定语从句和#非限定性定语从句。

答案

1 b. 2 谷歌目的在于安全运行网络（业务基础）

翻译

Google's Elite Security Team, Project Zero

谷歌黑客精英

When Apple launched the iPhone / through an exclusive partnership / with AT&T / in 2007, /
苹果发行 iPhone/ 通过独家合作伙伴的方式 / 和 AT&T/ 在 2007 年，/
seventeen-year-old George Hotz wanted to use an iPhone /
17 岁的乔治・霍兹想用 iPhone/
but not with AT&T. / He wanted to make calls /
但不通过 AT&T。/ 他想实现通话 /
through his own T-Mobile network, / so he cracked AT&T's lock / on the iPhone [1]. /
用他自己的 T-mobile 网络。/ 于是他破解了 AT&T 的锁 / 在 iPhone 里的。/
Apple scrambled to fix the bug / that allowed this, / but officially ignored George Hotz. /
苹果迅速修复了 Bug/ 引发上述状况的，/ 但官方无视了乔治・霍兹。/

Later in 2011, / Hotz reverse engineered the Playstation 3 / and posted a copy of the root keys /
2011 年下半年，/ 霍兹逆向破解了 PS 3/ 并将根密钥副本上传 /
on his website. / Sony sued him / but settled / after Hotz promised /
到他的网站。/ 索尼起诉了霍兹 / 但最后双方和解 / 在霍兹保证 /
never to hack Sony products again. /
不再入侵索尼产品后。/

Then in early 2014, / at Google's hacking competition, / Hotz discovered a security hole /
之后，到 2014 年初，/ 在谷歌的黑客攻击大赛上 / 霍兹发现了安全漏洞 /
in Google's Chrome OS. / The company gave him a $150,000 reward. / Two months later, /
在谷歌 Chrome OS 中。/ 公司给了霍兹 15 万美元奖励。/ 两个月后，/
Google's security engineer Chris Evans offered him / a position / in a team of elite hackers. /
谷歌的安全工程师克里斯・埃文斯提供给他 / 一个职位 / 在精英黑客团队。/
George Hotz accepted the offer / and now works / for Google's security team *Project Zero* [2]. /
霍兹接受了这个职位 / 现在工作 / 为谷歌的安全团队 Project Zero。/

Project Zero worked / in secret / until Google publically revealed the team / in July 2014. /
Project Zero 运作 / 秘密地 / 直到谷歌对外公布这个团队 / 在 2014 年 7 月。/
Its sole mission is tracking down / and getting rid of security flaws / in the world's software. /
它唯一的工作就是追踪 / 并解除安全漏洞 / 在全球软件中。/
These flaws are called *zero-day* vulnerabilities, / which are a common target / of cyber criminals. /
这些漏洞被称为“零日漏洞”，/ 它们是常攻击的目标 / 网络罪犯。/

Project Zero's hackers aren't just looking into the products / that Google makes. /
Project Zero 的黑客并不仅调查产品 / 谷歌制造的。/
They are free to hack any software / in the world. / Why? / They want to make a safer Internet /
他们自由入侵任何软件 / 世界上的。/ 为什么？ / 他们想构建一个更安全的网络 /
for everyone. / The team's policy is simple. / The team notifies vendors / of vulnerabilities / immediately. /
为所有人。/ 这个团队的策略很简单。/ 他们通知厂商 / 漏洞 / 立即。/

If fixes are not available / within 90 days, / bug reports automatically become available / to the public. /
如果没有修复 / 在 90 天内，/ 漏洞报告自动公布 / 给大众。/
The 90-day disclosure policy appears to be working / in most cases. /
90 日自动公布政策似乎很有效果 / 在多数情况下。/
The Adobe Flash team fixed 37 Project Zero vulnerabilities (or 100%) / within the 90-day period. /
Adobe Flash 团队修复了 37 个 Project Zero 漏洞（100%）/ 在 90 天内。/
The Project Zero blog indicates / that 85% of all vulnerabilities are patched / before the deadline [3]. /
Project Zero 的博客显示 /85% 的漏洞得到修补 / 在截止日期前。/

However, / recently / Google's strict 90-day policy came under fire / from Microsoft and Apple. /
但是，/ 最近 / 谷歌严苛的 90 日规定备受非议 / 被微软和苹果。/
The Project Zero team publicly disclosed bugs / which were present in Windows 8.1 and MacOS X /
Project Zero 团队将漏洞公之于众 / 它们存在于 Windows 8.1 和 Mac OS X 中 /
before Microsoft and Apple released patches. / Microsoft heavily criticized Google /
在微软和苹果发布补丁之前。/ 微软强烈谴责了谷歌 /
since the company was scheduled / to release a patch / just two days later [4]. / Recently /
因为微软已经计划 / 发布补丁 / 仅在 2 天后。/ 最近 /
Google loosened its 90-day policy / with an additional 14-day grace period. /
谷歌放宽了 90 日规定政策 / 另行宽限 14 天。/
Now vendors have an additional 14 days / to patch vulnerabilities / as long as they inform Google /
现在企业额外有 14 天 / 修复漏洞 / 只要它们通知谷歌 /
of the release schedule / before the deadline. /
发布日程 / 在截止日期前。/

"People deserve to use the Internet / without fear / that vulnerabilities out there / can ruin their privacy /
"人们有权使用互联网 / 不用害怕 / 网站中的漏洞 / 会毁掉他们的隐私 /
with a single website visit. /
因为浏览了一次网站。/
We're going to try to focus on the supply of these high value vulnerabilities /
我们将尽力锁定这些高价值漏洞 /
and eliminate them." / says Evans. /
然后排除。" / 埃文斯说。/

你的系统安全吗?

Zero-Day Attack: Is Your System Safe?

Last year Google revealed Project Zero, an elite team of top hackers. The team aims to improve security and protect the Internet from zero-day vulnerabilities. It has already uncovered a number of zero-day vulnerabilities in Microsoft Windows and Apple MacOS X [1].

But what exactly does *zero-day* mean? Before we dig into the meaning of *zero-day*, it's important to understand the difference between vulnerability and exploit [2].

Vulnerability and Exploit

A *vulnerability* refers to a flaw in a system, device, or application. It could be a bug or a design error. The presence of a vulnerability, though, does not cause harm. It is just a state of being vulnerable.

On the other hand, an *exploit* can be really dangerous. Exploiting is the act of abusing or taking advantage of a vulnerability. Attackers exploit vulnerabilities and gain access to a system by using *exploit code*. The term *exploit code* is often shortened to just *exploit*.

Simply put, a *vulnerability* is a hole in a system while an *exploit* is code to break into the system through the hole [3].

Zero-day Attack

Now that we understand vulnerability and exploit, it's time to ask: What is zero-day?

The term *zero-day* refers to the number of days between the public disclosure of a vulnerability and exploitation of the vulnerability. It's called *zero-day* because developers have zero days to address the vulnerability before attackers start exploiting it. In short, *zero-day* means the problem is not fixed yet, and a *zero-day exploit* is code to hack into a system through a *zero-day vulnerability*. *Zero-day attack* means attacking a *zero-day* vulnerability.

Protecting yourself

Due to the very nature of zero-day exploits, no network is 100 percent safe. However, there are measures to take to lower the risk [4].

For individuals, a commonsense approach is essential. Never click on suspicious links in emails, Kakao Talk, Facebook or Twitter postings. Never open an email attachment from an unknown source. Always use caution when you download something from the Internet.

Businesses and organizations can also establish certain security procedures to ensure the safety of their networks. Educate employees on best security practices. Use virtual LANs to protect business-critical information. Implement an intrusion detection system to detect zero-day attacks. Don't forget to lock down wireless access points to prevent wireless attacks.

出处

1. Project Zero, http://goo.gl/voRL7n
2. Zero-day attack, Wikipedia, http://goo.gl/wup7
3. The difference between an expoit and vulnerability http://goo.gl/hlz2uW
4. What is a Zero-Day Exploit? http://goo.gl/W5W4gK

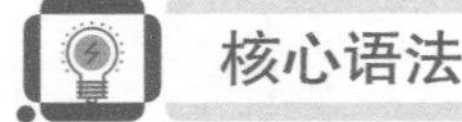

核心语法

熟悉 it 的多种用法。例文中将介绍 # 代名词、# 形式主语、# 非人称代词作主语的句子。# 形式主语并不是真正的主语，真正的主语由于过长而放在后半句。# 非人称代词指天气、时间、距离等，不再赘述。

- The team aims to protect internet users. **It** has already uncovered a number of bugs. 团队的目标是/保护互联网用户。/它（团队）已经发现了/大量Bug。
- To understand the difference is important. 理解不同很重要。
  = **It**'s important to understand the difference.
- **It**'s time to ask a question. 现在是提问时间。

熟悉 # 不定式 to 作 # 名词、# 形容词、# 副词的句子。从最后一个例句可以看出，to 具备众多含义。根据上下文理解即可。

- The team aims **to improve** security. 团队的目标/增进安全。
- Don't forget **to lock** the door. 不要忘记/锁门。
- An exploit is a code **to break** into the system. 利用漏洞的攻击是一种代码/入侵系统。
- There are measures **to take**. 有措施/要采取的（措施）。
- There are measures to take **to lower** the risk. 有要采取的措施/为了降低风险。
- Don't forget to lock the door **to prevent** intrusion. 要记得锁门/以避免入侵。
- Implement an intrusion detection system **to detect** zero-day attacks. 实施防入侵系统/阻止“零日攻击”的。或者实施防入侵系统/以阻止“零日攻击”。

熟悉使用 # 动名词的句子。在 # 动词原形后加 ~ing 用作 # 名词，所以称为 # 动名词。可以理解为“做某事”。

- Attackers start **exploiting** it. 攻击人士开始/恶意利用它。
- Zero–day attack means **attacking** a zero–day vulnerability. “零日攻击”意味着/攻击“零日漏洞”。

熟悉 # 连接副词，必须熟记含义。

- The presence of a vulnerability, **though**, does not cause harm. 漏洞的存在，然而，并没有带来危害。
- **On the other hand**, an exploit can be really dangerous. 另一方面，利用漏洞的攻击是相当危险的。
- **Simply put**, a vulnerability is a hole in a system. 简言之，漏洞就是个洞/系统中的。
- **In short**, zero–day means (that) the problem is not fixed yet. 简言之，“零日”意味着/问题尚未解决。

单词&短语

reveal 揭露，泄漏
aim to V ~ 目的，旨在 ~，致力于 ~
improve 改善，增进
vulnerability 漏洞
uncover 揭露，揭开盖子
a number of 大量（=many）
exactly 准确地
dig into 挖掘，探究
difference 不同点
exploit 压榨，滥用，攻击程序
refer to 指的是
flaw 缺点，瑕疵
presence 存在
cause 诱发，引起
harm 危害

state 状态
on the other hand 另一方面
abuse 滥用，误用
take advantage of 利用
attacker 攻击者
gain access to ~ 接近 ~
shorten 缩短
simply put 简言之
it's time to V 到 ~ 的时间了
the number of ~ ~ 的数量
public disclosure 公布
exploitation 剥削，滥用
address 处理
in short 简言之
due to ~ 由于，因为 ~

nature 本性
take measures 采取措施
lower 降低
commonsense 常识性的
approach 接近方法，途径
essential 必要的
suspicious 可疑的
email attachment 附件
unknown source 出处不明
organization 组织，团体
establish 设立，创立
security procedure 安全程序
ensure 保证
safety practice 安全措施
virtual LAN 虚拟局域网
business-critical 商业机要的
implement 实施
intrusion detection 入侵检测
detect 检查，侦测
wireless access point 无线接入点

技术术语

vulnerability window 漏洞窗口是指，从首次恶意利用漏洞到为应对危机而研发并运行补丁的时间段。

- 研发人员开发包括不为人知的漏洞在内的软件。
- 攻击者比研发人员更早找到漏洞。
- 攻击者编写利用漏洞的攻击代码。
- 研发人员或者外部团体发现可能被攻击的漏洞后，研发人员开始研发补丁。
- 研发人员发布补丁。

研发人员开发 / 发布补丁之前，攻击者抢先攻击的情况称为“零日攻击”。因为研发人员无法得知攻击者何时发现漏洞，所以漏洞窗口的时长可能持续几个月或者几年。

根据提示完成句子

A ____________ ______ __ _ ____ / in a system, device,
漏洞指缺陷 /存在于系统、设备、

or application. / It could be a bug /
应用中的。 /它有可能是Bug/

or a design error. / The ________ __ _ ______________, / though, /
或设计错误。 /漏洞的存在， /然而，/

does not _____ ____. /
并没有害处。

It is just a _____ / of _____ __________. /
它只是一种状态 /易受攻击的。/

__ ___ _____ ____, / an _______ can be really __________. /
另一方面， /利用漏洞的攻击非常危险。/

__________ is the ___ /
利用漏洞进行攻击是一种行为/

__ ________ / or ______ __________ __ _ ____________. /
滥用 /或者利用漏洞。/

_________ _______ _______________ /
攻击者借助漏洞进行攻击/

and ____ ______ / to a system / by using ________ ____. /
并且进入 /系统/ 借助利用漏洞攻击的代码。/

The term ________ ____ /
利用漏洞攻击的代码/

is often _________ / to just _______. /
经常被缩写为/ 表示利用漏洞的攻击。/

______ ___, / a ______________ is a ____ / in a system /
简言之，/ 漏洞就是个洞/ 系统中的/

while an __________ is ____ / to _____ ____ the system /
然而利用漏洞的攻击是代码/ 以入侵系统/

________ ___ ____. /
通过漏洞。/

___ ____ we understand ______________ and _______, /
现在大家已经理解漏洞和利用漏洞的攻击，/

____ time __ ask: /
下面是提问时间：/

What is _________? /
何为“零日”？/

思考题

解答题

1 （理解）以下哪一项预防“零日攻击”的方式比较切合实际？

ⓐ 切断网线　　ⓑ 不随意从网上下载程序

ⓒ 禁止使用无线 LAN　　ⓓ 禁止使用 SNS

2 （理解）以下哪一项与例文叙述的“零日攻击”不相符？

ⓐ“零日攻击”与发现漏洞并利用漏洞攻击的节点一致。

ⓑ“零日攻击”没有给研发人员留出修复漏洞的时间。

ⓒ“零日攻击”无法预防。

ⓓ“零日攻击”主要通过网络实现。

3 （论述）在无法继续获得支持的操作系统中，“零日攻击”危害性更大的原因是什么？

4 （论述）列举个人应对“零日攻击”的方法。

讨论

1 在全球互联网时代，请简述摆脱“零日攻击”的方法。

2 漏洞和利用漏洞的攻击相比，哪一种更容易阻止？

课堂总结

1 学习it作#形式主语和#非人称主语的句子。

2 学习#动名词的概念和例句。了解具有类似形态的#现在分词。

3 熟记常见的#连接副词。首先熟记例文中出现的内容。后文也将陆续出现。

答案

1 b. 2 c. 3 因为不是在操作系统级别修复漏洞，所以利用漏洞的攻击会变为众多形态，很难检测和防御。4 不点击SNS上的可疑链接，不下载邮件的可疑附件，从网上下载要时刻保持警惕。

翻译

Zero-Day Attack: Is Your System Safe?

你的系统安全吗?

Last year / Google revealed Project Zero, / an elite team of top hackers.

去年 / 谷歌揭示了 Project Zero，/ 一个顶级黑客精英团队。/

The team aims / to improve security / and protect the Internet / from zero-day vulnerabilities. /

这个团队的目标是 / 改善安全 / 保护互联网 / 免受“零日漏洞”的危害。/

It has already uncovered / a number of zero-day vulnerabilities /

他们已经发现了 / 大量“零日漏洞”/

in Microsoft Windows and Apple MacOS X [1]. /

在微软 Windows 和 Apple Mac OS X 中。/

But what exactly does zero-day mean? / Before we dig into the meaning of zero-day, /

但是“零日”的准确内涵是什么呢? / 在我们探究其涵义之前，/

it's important to understand the difference / between vulnerability and exploit [2]. /

重要的是了解不同点 / 存在于漏洞和漏洞利用之间的。/

Vulnerability and Exploit / 漏洞和漏洞利用

A vulnerability refers to a flaw / in a system, device, or application. / It could be a bug /

漏洞指缺陷 / 存在于系统、设备、应用中的。/ 它有可能是 Bug/

or a design error. / The presence of a vulnerability, / though, / does not cause harm. /

或设计错误。/ 漏洞的存在，/ 然而，/ 并没有害处。/

It is just a state / of being vulnerable. /

它只是一种状态 / 易受到攻击的。/

On the other hand, / an exploit can be really dangerous. / Exploiting is the act /

另一方面，/ 漏洞利用非常危险。/ 漏洞利用是一种行为 /

of abusing / or taking advantage of a vulnerability. / Attackers exploit vulnerabilities /

滥用 / 或者利用漏洞。/ 攻击者利用漏洞 /

and gain access / to a system / by using exploit code. / The term exploit code /

并且进入 / 系统 / 借助漏洞利用代码。/ 术语漏洞利用代码 /

is often shortened / to just exploit. /

经常缩写为 / 漏洞利用。/

Simply put, / a vulnerability is a hole / in a system / while an exploit is code /

简言之，/ 漏洞就是个洞 / 系统中的 / 而漏洞利用是代码 /

to break into the system / through the hole [3]. /

以入侵系统 / 通过漏洞。/

Zero-day Attack / “零日攻击”

Now that we understand vulnerability and exploit, / it’s time to ask: / What is zero-day? /
现在大家已经理解漏洞和漏洞利用代码，/ 下面是提问时间: / 何为“零日”？ /

The term zero-day refers to the number of days / between / the public disclosure of a vulnerability /
术语“零日”指天数 / 介于 / 公开发布漏洞和 /
and exploitation of the vulnerability. / It’s called zero-day /
漏洞利用。/ 这就是所谓的“零日”/
because developers have zero days / to address the vulnerability /
因为研发人员剩下“零日”/ 处理漏洞 /
before attackers start exploiting it. / In short, / zero-day means / the problem is not fixed yet, /
在攻击者利用漏洞进行攻击前。/ 简言之，/“零日”意味着 / 问题尚未解决，/
and a zero-day exploit is code / to hack into a system / through a zero-day vulnerability.
利用“零日漏洞”的攻击是代码 / 以入侵系统 / 通过“零日漏洞”。/
Zero-day attack means / attacking a zero-day vulnerability. /
“零日攻击”意味着 / 攻击“零日漏洞”。/

Protecting yourself / 自我保护

Due to the very nature / of zero-day exploits, / no network is 100 percent safe. /
因为本质 /“零日漏洞攻击”的，/ 没有 100% 安全的网络。/
However, there are measures / to take to lower the risk [4]. /
但是，可以采取措施 / 降低风险。/

For individuals, / a commonsense approach is essential. / Never click on suspicious links /
对个人而言，/ 采取常识性的访问方式是必须的。/ 勿点击可疑链接 /
in emails, Kakao Talk, Facebook or Twitter postings. / Never open an email attachment /
在邮箱、Kakao Talk、Facebook 或者 Twitter 推文中的。/ 勿打开邮箱附件 /
from an unknown source. / Always use caution / when you download something /
来源不明的。/ 时刻保持警惕 / 下载的时候 /
from the Internet. /
从网络上。/

Businesses and organizations can also establish certain security procedures /
公司和组织也可以设置特定的安全程序 /
to ensure the safety / of their networks. / Educate employees / on best security practices. /
以确保安全 / 他们网络的。/ 对员工进行培训 / 用最好的安全程序。/
Use virtual LANs / to protect business-critical information. / Implement an intrusion detection system /
使用虚拟局域网 / 保护商业机要信息。/ 实施入侵探测系统 /
to detect zero-day attacks. / Don’t forget to lock down wireless access points /
以探测“零日攻击”。/ 切记关掉无线接入点 /
to prevent wireless attacks. /
以避免无线攻击。

我的联想笔记本也不安全吗？

Lenovo Superfish Scandal

Lenovo's Superfish fiasco might be the worst security scandal of the year, maybe even of the decade. It turned out that the company shipped laptops with pre-installed adware that would open a security hole. The adware called *Superfish* would leave millions of laptops vulnerable to cyberattacks.

Is Superfish malware?

Adware is a software application that automatically displays advertisements. Superfish is an adware program that inserts advertisements into web pages users visit. Generally, adware is annoying but not really malware. However, security experts consider Superfish to be malware, even an especially bad kind.

Superfish used fake, self-signed root certificates to hijack HTTPS traffic and inject its own ads onto websites. What does that mean? Errata Security's Robert Graham, who reverse engineered the Lenovo/Superfish certificate and cracked the password, explains its implication in simple terms:

> "[Superfish] is designed to intercept all encrypted connections, things it shouldn't be able to see. It does this in a poor way that it leaves the system open to hackers or NSA-style spies. For example, it can spy on your private bank connections..."[1]

Who is affected?

Lenovo has stopped pre-installing Superfish since January 2015. According to the company's spokesman, Lenovo installed Superfish on consumer PCs and laptops only from September to December 2014. Lenovo shipped a total of 16 million units over the period. Chrome, Internet Explorer, and Firefox are all affected.

Am I infected?

If you have purchased a Lenovo PC or laptop within the last two years, it would be wise to check if your system came with Superfish. First, you might want to check out Lenovo's Superfish Vulnerability page which lists affected models [2].

LastPass has created a web tool to quickly verify if your computer is infected. In Internet Explorer or Chrome, simply visit the site [3]. If you do not see a "You are Safe" message, you are infected.

How to remove Superfish?

Manual instructions are now available from many trusted sources [1, 4]. On its own support site, Lenovo has published an automatic removal tool as well as manual removal instructions [2]. Microsoft and McAfee also updated their antivirus software to detect and remove Superfish [5]. So if you simply run a full system scan with your legitimate antivirus software, you should be free from Superfish.

出 处

1. Lenovo responds to "Superfish" report, says malware is no longer active, http://goo.gl/ckoRvt
2. Superfish Vulnerability, http://goo.gl/TxnuDS
3. LastPass Superfish Checker, https://goo.gl/cAf8vN
4. How to Test Your PC for the New "Superfish" Security Vulnerability, http://goo.gl/rdcayl
5. Microsoft, McAfee update antivirus software to protect against Superfish, http://goo.gl/DuuP6P

核心语法

熟悉 # 现在完成时态〈have/has+ 过去分词〉。汉语中用“已经、过、了”等表示，指过去发生的事情对现在造成的影响。

- Lenovo **has stopped** pre-installing Superfish since January 2015. 联想已经停止 / 预装Superfish /从2015年1月开始。（=现在不再安装）
- If you **have purchased** a Lenovo PC within the last two years, ~如果你在过去2年内购买了联想PC，~（=如果现在有）
- Lenovo **has published** an automatic removal tool on its website. 联想已经发布了/自动删除工具/在其网站上。（=现在在网站上）

熟悉 # 关系代名词。# 关系代名词连接两个句子，兼具 # 连词和 # 代名词的作用。关系代名词前加（,）的 # 非限定性定语从句用于补充说明 # 先行词。

- Adware is **a software application**. + **It** automatically displays ads. 广告软件是软件应用。+ 它自动展示广告。
  = Adware is **a software application** (**that** automatically displays ads). =广告软件是软件应用/自动展示广告的。
- The company shipped laptops with **adware** (**that** opens a security hole). 公司在电脑上安装了/广告软件/打开安全漏洞的。
- **Robert**, (**who** cracked the password,) explains its implication. 罗伯特，他破解了密码，对它做出了解释。

熟悉 # 关系代名词的省略。# 宾格关系代名词、# 主格关系代名词 +be 动词可以省略。省略后不影响原意。

- Superfish is **an adware program that** inserts advertisements into **web pages** ([**that**] users visit). Superfish是广告软件程序/插入广告的/在网页/用户浏览的。

- The **adware** ([**which** is] called Superfish) leaves millions of laptops vulnerable to cyberattacks. 广告软件/被称为Superfish的/让众多笔记本电脑处于易受网络攻击的状态。

熟悉 #if 引导的两种句子。if 表“是否”时，可以用 whether 代替。

- **If** you do not see a “You are Safe” message, you are infected. 如果没有看到“你是安全的”的信息 / 那么你的电脑被感染了。
- LastPass has created a web tool to quickly verify **if/whether** your computer is infected. LastPass创建了网页工具/迅速验证的/你的电脑是否被感染。

单词&短语

fiasco 惨败
security 安全
scandal 绯闻
turn out ~ 被证明是 ~
ship 出货
pre-installed 预装
vulnerable to ~ 易受 ~ 攻击的
automatically 自动地
advertisements 广告
insert a into b 将 a 插入 b
generally 一般地
fake 虚假的
root certificate 根证书
hijack 劫持，拦截
inject 注入
reverse engineer 逆向工程
crack a password 破解密码
explain 解释
implication 含义
in simple terms 用简单的术语
be designed to V 设计目的为 ~
intercept 拦截
encrypted connection 加密连线
spokesman 发言人
over the period 在此期间
affect 影响
infect 感染
purchase 购买
remove 删除
manual instruction 手册
vulnerability 漏洞
verify 证明，确认

技术术语

man in the middle attack 中间人攻击是指，通过拦截正常的网络通信数据进行数据篡改或转移，而通信双方毫不知情。Superfish 也是一种利用中间人攻击的恶意软件。

根据提示完成句子

______ is a software application /
广告软件是一款软件应用/

that automatically displays ____________. /
它自动展示广告。/

Superfish is an ______ program / that inserts ____________ /
Superfish是广告软件程序/ 插入广告的/

into web pages / users visit. /
在网页/ 用户访问的。/

Generally, / ______ is _________ / but not really _______. /
一般来说，/ 广告软件很让人厌烦，/ 但它不是恶意软件。/

However, / ________ ________ consider Superfish / to be _______, /
但是，/ 安全专家认为Superfish/ 是恶意软件，/

even an especially ___ ____. /
甚至是特别恶意的。/

Superfish used ____, ____-______ ____ ____________ /
Superfish使用假的、自己署名的根证书/

to _______ HTTPS traffic / and ______ its own ads / onto websites. /
以拦截HTTPS通信/ 并将自己的广告植入/ 网站。/

思考题

解答题

1 （理解）以下哪项准确描述了Superfish恶意软件的特性？

ⓐ 利用浏览器的“零日漏洞”。

ⓑ 不区分型号 / 样式，安装在 Windows 系统所有台式机 / 笔记本电脑中。

ⓒ 销售时预装的软件。

ⓓ 目前无法修复。

2 （理解）以下哪一项是Superfish的初衷？

ⓐ 搜索并杀毒

ⓑ 向用户展示广告

ⓒ 自动进行系统更新

ⓓ 在浏览器拦截用户不愿意看到的广告

3 （论述）请列出Superfish 恶意软件存在危险性的最主要原因。

4 （论述）请列出降低Superfish危险性的方法。

讨论

1 Superfish与现存的恶意软件传播方式相比，有哪些不同？从企业角度看，应该采取何种措施避免今后发生类似事件？

2 调查Superfish使用的自主署名认证书能引起严重安全问题的技术层面原因。

课堂总结

1 学习#现在完成时态和例句。

2 学习#限定性定语从句和#非限定性定语从句。

3 学习#关系代名词的省略。

答案

1 c. 2 b. 3 用TLS（HTTPS）可以拦截加密通信，所以如果有其他恶意软件入侵，那么只要控制Superfish，个人敏感信息（银行账户信息）就很可能泄漏。 4 自动/手动删除Superfish。利用杀毒软件修复。

翻译

Lenovo Superfish Scandal

我的联想笔记本也不安全吗?

Lenovo's Superfish fiasco might be the worst security scandal / of the year, /
联想的 Superfish 惨败可能是最糟的安全传言 / 这一年的，/
maybe even of the decade. /
甚至可能是近 10 年的。/
It turned out / that the company shipped laptops / with pre-installed adware /
事实证明 / 公司在笔记本电脑出货时 / 预装广告软件 /
that would open a security hole. / The adware / called Superfish /
开启安全漏洞的 / 这个广告软件 / 称为 Superfish/
would leave millions of laptops vulnerable / to cyberattacks. /
将使数百万台笔记本电脑处于易受攻击状态 / 来自网络的。/

Is Superfish malware? / Superfish是危险软件吗?

Adware is a software application / that automatically displays advertisements. /
广告软件是一款软件应用 / 它自动展示广告。/
Superfish is an adware program / that inserts advertisements / into web pages /
Superfish 是广告软件程序 / 插入广告的 / 在网页 /
users visit. / Generally, / adware is annoying / but not really malware. / However, /
用户访问的。/ 一般来说，/ 广告软件很让人厌烦，/ 但它不是恶意软件。/ 但是，/
security experts consider Superfish / to be malware, / even an especially bad kind. /
安全专家认为 Superfish/ 是恶意软件，/ 甚至是特别恶意的软件。/

Superfish used fake, self-signed root certificates / to hijack HTTPS traffic /
Superfish 使用假的、自己署名的根证书 / 以拦截 HTTPS 通信 /
and inject its own ads / onto websites. / What does that mean? /
并将自己的广告植入 / 网站。/ 这意味着什么呢? /
Errata Security's Robert Graham, / who reverse engineered the Lenovo/Superfish certificate /
Errata Security 的罗伯特·格拉哈姆，/ 逆向分析了联想 /Superfish 证书 /
and cracked the password, / explains its implication / in simple terms: /
并且破解了密码，/ 解释了其含义 / 用简单术语: /

> "[Superfish] is designed / to intercept all encrypted connections, things /
> "Superfish 被设计 / 为拦截所有加密连接和通信内容 /
> it shouldn't be able to see. / It does this / in a poor way /
> 它不应当看到的/它这样做/用糟糕的方式/
> that it leaves the system open / to hackers or NSA-style spies. / For example, /
> 使系统暴露/给黑客或者NSA-style的间谍。/例如，/

it can spy / on your private bank connections..."[1] /
它能入侵/你的个人网银……” /

Who is affected? / 谁被影响?

Lenovo has stopped / pre-installing Superfish / since January 2015. /
联想已经停止 / 预装 Superfish/ 从 2015 年 1 月。/
According to the company's spokesman, / Lenovo installed Superfish / on consumer PCs and laptops /
据公司发言人称，/ 联想安装 Superfish/ 在消费者的 PC 和笔记本电脑中 /
only from September / to December / 2014. / Lenovo shipped a total of 16 million units /
仅从 9 月 / 到 12 月 /2014 年。/ 联想共出货 1600 万台 /
over the period. / Chrome, Internet Explorer, and Firefox are all affected. /
在此期间。/Chrome、Internet Explorer、火狐全部受到影响。/

Am I infected? / 我也被影响了吗?

If you have purchased a Lenovo PC or laptop / within the last two years, /
如果已经购买了联想 PC 或者笔记本电脑 / 在过去 2 年内，/
it would be wise to check / if your system came with Superfish. / First, /
检查一下是很明智的 / 你的系统是否安装 Superfish。/ 首先，/
you might want to check out Lenovo's Superfish Vulnerability page /
你可能会检查联想电脑的 Superfish 漏洞页面 /
which lists affected models [2]. /
那里列出了受影响的型号。/

LastPass has created a web tool / to quickly verify / if your computer is infected. /
LastPass 制作了一个网页工具 / 迅速检查的 / 你的电脑是否已经被感染。/
In Internet Explorer or Chrome, / simply visit the site [3]. /
用 Internet Explorer 或者 Chrome，/ 访问网站。/
If you do not see a "You are Safe" message, / you are infected. /
如果没有看到“你是安全的”的信息，/ 那么你的电脑被感染了。/

How to remove Superfish / 如何删除**Superfish**?

Manual instructions are now available / from many trusted sources [1, 4]. /
现在可以使用手册 / 来自许多值得信赖的出处。/
On its own support site, / Lenovo has published an automatic removal tool /
在它自己的网站，/ 联想已经发布了一款自动删除工具 /
as well as manual removal instructions [2]. /
同时也发布了手动删除指南。/
Microsoft and McAfee also updated their antivirus software /
微软和 McAfee 也更新了他们的杀毒软件 /
to detect and remove Superfish [5]. / So if you simply run a full system scan /
以检测和删除 Superfish。/ 所以如果你仅想给整个系统做个体检 /
with your legitimate antivirus software, / you should be free / from Superfish. /
用合法的杀毒软件，/ 那你应该不受 /Superfish 的影响。/

需要立即变更4种Facebook 设置

Four Facebook Settings You Should Change Now!

Facebook keeps adding new features. Some are useful, but some are annoying or may diminish your privacy. The features are optional, but Facebook often turns them on by default. The company probably hopes that you leave them on forever. Why?

Facebook makes money by selling your information to advertisers. More precisely, Facebook "collects all the data, divides it into categories, makes it anonymous, and sells it to advertisers in *buckets* of metadata."[1] Facebook claims that it is committed to protecting user privacy, but some argue that the company doesn't care about your privacy [2].

Either way, you have to stay on top of your own privacy settings. Check the following settings and make sure that your private information is protected [3].

Auto-play Video Ads

You scroll through your news feed, and videos start playing. You didn't click or tap the video, yet it plays automatically. Facebook launched this annoying feature in December 2013. Fortunately, you can turn it off easily.

First, click the down arrow in the upper-right corner of the page. Then select **Settings** in the drop-down menu. Click **Videos** in the left menu. Then change the **Auto-Play Videos** option to **Off**.

Default Privacy Settings

Facebook now offers the audience selector option, so you can adjust the audience for each individual post. However experts advise that you should always set the *default* audience properly.

First, navigate to your **Settings** page again. Click **Privacy** in the left column. The page will display a variety of privacy settings. Click **Edit** and change the default audience setting.

Search History

Facebook keeps all of your searches. Whenever you look for friends or posts, it gets saved to your search history. This supposedly helps your future searches, but you may not want to save every search. You can delete your entire search history with a couple of clicks.

Click the down arrow and select **Activity Log** from the drop-down menu. Menu options will be listed on the left side of the page. Click **More** under **Comments** to expand the list. Click **Search**, and all your searches will be displayed by date. Click **Clear Searches** at the top of the page.

Social Advertising

If you *like* a certain product or service on Facebook, Facebook will show its ad to your friends. Your friends will pay more attention to the ad *because you liked it*. It's called *social advertising*, and you may want to opt out of this feature.

Navigate to your **Settings** page. Click **Ads** on the left. Change both options from **Only Friends** to **No one**. Make sure to click **Save Changes**.

出 处

1. What Facebook knows about you, http://goo.gl/aigr0o
2. Facebook doesn't care about your privacy, that's why it is worth $150bn, http://goo.gl/DfHAns
3. Facebook's Help Center, https://goo.gl/siJoT

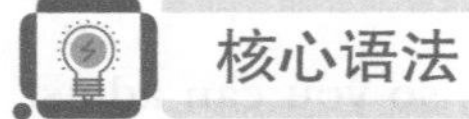

核心语法

熟悉 # 句式 3〈主语 + 动词 + 宾语〉中 # 不定式 to 和 # 动名词作宾语的句子。阅读和听力中，了解 # 不定式和 # 动名词可作 # 宾语即可；但在写作和会话中，需要熟记可接 # 不定式和 # 动名词作 # 宾语的动词。# 不定式、# 动名词作 # 名词时，译为“~做某件事”。

- Facebook keeps **adding** new features. Facebook持续/增加新功能。
- Videos start **playing**. 视频开始/播放。
- You may not want **to save** every search. 你可能不想/每次都保存搜索记录。

熟悉 # 句式 3〈主语 + 动词 + 宾语〉中 #that 从句作 # 宾语的句子。#that 引导的宾语从句中，that 作宾语经常被省略，例文已全部标出。

- The company hopes **that** you leave them on. 公司希望/你打开它们。
- Facebook claims **that** it is committed to protecting user privacy. Facebook主张/他们（Facebook）致力于保护用户隐私。
- Make sure **that** your private information is protected. 确保/你的私人信息得到保护。

熟悉软件菜单中的常用句型和 # 祈使句。# 祈使句省略主语 You，以 # 动词原形开头。意为“使、让”。

- **Click** the down arrow in the upper-right corner of the page. 请点击/下滑箭头/页面右上的。
- **Navigate** to your Settings page. 请找到/你的设置页面。
- **Change** both options from Only Friends to No one. 变更/两个选项/从仅限好友/到禁止任何人。

熟悉可连接两个句子的 # 连词。

- You didn't click or tap the video, **yet** it plays automatically. 你并没有点击或碰触视频，/但视频却自动播放。
- The company now offers the option, **so** you can adjust the setting. 公司现在设置了选项功能，/所以你可以改变设置。
- **Whenever** you look for something, it gets saved to your search history. 每当你搜索信息时，/这些信息都被保存到搜索记录。

单词&短语

keep V+ing 持续做 ~
annoy 让人厌烦
diminish 使减少
feature 功能
optional 可选的
turn on 打开
by default 默认
probably 可能
advertiser 广告商
more precisely 更准确地
collect 收集
divide A into B 将 A 分成 B
anonymous 匿名的
be committed to ~ing 致力于 ~、投身于 ~
protect 保护
care about ~ 关注 ~
either way 无论如何
stay on top of ~ 熟练掌握 ~
make sure 确保
auto-play 自动播放
scroll through 滚动
news feed 信息流
launch 上线
fortunately 幸运地
down arrow 下滑箭头
upper-right corner 右上角
drop-down menu 下拉菜单
change A to B 将 A 换成 B
offer 提供
audience 听众
adjust 调整
expert 专家
properly 适当地，正确地
navigate to ~ 浏览 ~
column 列
a variety of 多种多样的
search history 搜索记录
whenever 任何时候，每当 ~ 时
look for 寻找
supposedly 据推测，可能
entire search history 全部搜索记录
a couple of clicks 双击
expand 扩张
display by date 按照日期展示
pay attention to ~ 关注 ~
opt out of ~ 选择退出 ~

根据提示完成句子

Facebook keeps / ______ ___ ________. / Some are useful, /
Facebook持续/ 增加新功能。/ 有些是有用的，/

but some are annoying / or may _________ ____ ________. /
但有些却让人很厌烦/ 甚至可能减少你的隐私。/

The _______ ___ _________, / but Facebook _____ _____ _____ __ /
这些功能是可选的，/ 但Facebook经常将其打开/

__ _______. / The company probably hopes / that you _____ _____ __ /
默认。/ 公司可能希望/你将其打开/

_______. / Why? / Facebook _____ _____ / by _______ ____ __________ /
永远地。/ 为什么呢？/ Facebook盈利/ 通过出售你的信息/

to __________. / ____ __________, / Facebook "_______ ___ ___ ____, /
给广告商。/ 更确切地说，/ Facebook "将收集到的所有资料，/

_______ __ ____ __________, / _____ __ _________, / and _____ __ /
进行分类，/ 匿名处理后/ 出售给/

__ ___________ / in buckets of metadata." / Facebook claims /
广告商/ 大量元数据。" / Facebook主张/

that it is _________ / to __________ ____ ________, / but some argue /
他们致力于/ 保护用户隐私，/ 但有人认为/

that the company doesn't ____ _____ ____ _______. / _______ ___, /
公司根本不关注个人隐私。/ 无论如何，/

you have to ____ __ ___ __ / ____ ___ ________ _________. /
你都要更明确掌握/ 自己的隐私设置。/

Check the following settings / and ____ ____ /
检查以下设置/ 确保/

that your _______ ___________ is _________. /
你的个人信息得到保护。

思考题

解答题

1 （理解）以下哪一项为Facebook的盈利方式？

ⓐ 捐赠　　ⓑ 加盟费

ⓒ 电商收益　　ⓓ 出售个人信息及广告

2 （理解）以下各项中，哪一项没有准确说明例文阐述的Facebook隐私政策？

ⓐ 提供默认隐私变更选项。

ⓑ 无法关闭视频广告自动播放。

ⓒ Facebook 保存个人全部搜索记录。

ⓓ 社交广告影响好友。

3 （论述）请列举避免Facebook个人信息泄漏的注意事项。

4 （论述）请列举Facebook社交广告的优点。

讨论

1 你更倾向Facebook默认选项强化个人隐私保护，还是倾向让其更加顺畅地公开信息？

2 Facebook运营比较循规蹈矩，以避免内部信息泄漏（例如：搜索引擎）。请简述这种政策的优缺点。

课堂总结

1 学习#动词后接#不定式和#动名词作#宾语的例句。

2 学习#动词后接#that引导的宾语从句作#宾语的例句。

3 学习表允许、禁止、共动的#祈使句。

答案

1 d. 2 b. 3 使用默认隐私保护设置。删除搜索记录。关闭社交广告功能。 4 好友之间关注的事情比较类似，看到相关内容的可能性很大。

翻译

Four Facebook Settings / You Should Change Now!
需要立即变更4种Facebook设置

Facebook keeps / adding new features. / Some are useful, / but some are annoying /
Facebook 持续 / 增加新功能。/ 有些很实用，/ 但有些却让人很厌烦 /
or may diminish your privacy. /
甚至可能减少隐私。/
The features are optional, / but Facebook often turns them on / by default. /
这些功能是可选的，/ 但 Facebook 经常将其打开 / 默认。/
The company probably hopes / that you leave them on / forever. / Why? /
公司可能希望 / 你将其打开 / 永远地。/ 为什么呢？ /

Facebook makes money / by selling your information / to advertisers. / More precisely, /
Facebook 盈利 / 通过出售你的信息 / 给广告商。/ 更确切地说，/
Facebook "collects all the data, / divides it into categories, / makes it anonymous, /
Facebook "将收集到的所有资料，/ 进行分类，/ 匿名处理后，/
and sells it / to advertisers / in buckets of metadata."[1] / Facebook claims /
出售 / 给广告商 / 大量元数据。" /Facebook 主张 /
that it is committed / to protecting user privacy, / but some argue /
他们致力于 / 保护用户隐私，/ 但有人认为 /
that the company doesn't care about your privacy [2]. /
公司根本不关注个人隐私。/
Either way, / you have to stay on top of / your own privacy settings. /
无论如何，/ 你都要更了解 / 自己的隐私设置。/
Check the following settings / and make sure / that your private information is protected [3]. /
检查以下设置 / 确保 / 你的个人信息得到保护。/

Auto-play Video Ads / 自动播放视频广告

You scroll through your news feed, / and videos start playing. / You didn't click /
你滚动信息，/ 视频开始播放。/ 你并没有点击 /
or tap the video, / yet it plays automatically. / Facebook launched this annoying feature /
或碰触视频，/ 但视频却自动播放。/Facebook 上线了这种让人厌烦的功能 /
in December 2013. / Fortunately, / you can turn it off / easily. /
在 2013 年 12 月。/ 幸运的是，/ 你可以关掉这个功能 / 轻而易举地。/
First, / click the down arrow / in the upper-right corner of the page. /
首先，/ 点击下拉箭头 / 页面右上端的。/
Then select Settings / in the drop-down menu. / Click Videos / in the left menu. /
然后选择设置页面 / 在下拉菜单。/ 点击视频 / 左侧菜单中的。/
Then change the Auto-Play Videos option / to Off. /
然后将自动播放视频按钮 / 调整为 "关"。/

Default Privacy Settings / 默认隐私设置

Facebook now offers the audience selector option, / so you can adjust the audience /
Facebook 支持受众选择功能。/ 因此你可以选择用户 /
for each individual post. / However experts advise /
为每个帖子。/ 但专家建议 /
that you should always set the default audience / properly. /
你应该经常设置默认受众 / 正确地。/
First, navigate to your Settings page again. / Click Privacy / in the left column. /
首先，再次进入设置页面。/ 点击"隐私"/ 在左栏中。/
The page will display a variety of privacy settings. / Click Edit /
页面将展示多种隐私设置。/ 点击"编辑"/
and change the default audience setting. /
改变默认受众设置。/

Search History / 搜索记录

Facebook keeps / all of your searches. / Whenever you look for friends or posts, /
Facebook 保存 / 所有搜索记录。/ 每当你搜索朋友或者搜索帖子时，/
it gets saved to your search history. / This supposedly helps your future searches, /
它们都被保存到搜索记录。/ 这可能有利于今后搜索，/
but you may not want / to save every search. / You can delete your entire search history /
但可能你并不喜欢 / 保存所有搜索记录。/ 你可以删掉所有搜索记录 /
with a couple of clicks. /
通过双击。/
Click the down arrow / and select Activity Log / from the drop-down menu. /
点击下滑箭头 / 选择"活动日志"/ 在下拉菜单。/
Menu options will be listed / on the left side / of the page. /
菜单选项会被列在 / 左侧 / 页面的。/
Click More under Comments / to expand the list. / Click Search, /
点击"说明"下端的"更多"/ 展开目录。/ 点击"搜索"/
and all your searches will be displayed / by date. / Click Clear Searches / at the top of the page. /
所有搜索结果都被展示出来 / 按照日期。/ 点击"清除搜索"/ 页面顶端的。/

Social Advertising / 社交广告

If you like a certain product or service / on Facebook, / Facebook will show its ad /
如果你点赞了特定商品或者服务 / 在 Facebook，/Facebook 将会展示相关广告 /
to your friends. / Your friends will pay more attention / to the ad / because you liked it. /
给你的朋友。/ 你的朋友会更关注 / 这些广告 / 因为你喜欢。/
It's called social advertising, / and you may want to opt out of / this feature. /
这就是所谓的社交广告，/ 可能你想退出 / 这个功能。/
Navigate to your Settings page. / Click Ads on the left. / Change both options /
搜索"设置"页面 / 点击左侧"广告"/ 改变两个选项 /
from Only Friends / to No one. / Make sure / to click Save Changes. /
从"仅对好友可见"/ 到"不允许任何人查看"。/ 确保 / 点击"保存更改"。/

病毒与恶意软件区别何在？

Viruses and Malware: What's the Difference?

Many PC users consider viruses, malware, spyware, adware, worms, and Trojans all to be the same thing. While all of these are malicious programs that can harm our computers, each behaves differently.

Malware is a combination of two words: 'malicious' and 'software'. As the term suggests, any program whose intent is malicious is malware. That is, any hostile or intrusive programs, including viruses, spyware, adware, worms, and Trojans, are all malware.

Malware is classified into several types based on several criteria. The most commonly used criterion is infection method. Depending on how it infects a system, malware generally fits into one of the following categories:

Virus

A virus is malware that attaches itself to (or infects) other programs or files. When a user runs an infected program (or a host program), the virus runs also and performs malicious activities. It also reproduces itself by attaching itself to additional files. Viruses spread through human actions such as forwarding an infected file, opening an infected email attachment, or downloading and running a malicious program from the Internet.

Worms

Worms are malware that run by themselves. They are self-replicating and self-propagating, which means they do not need host programs or human actions to spread. Worms travel through networks on their own while replicating themselves. They can spread very fast because there is no need for human interaction.

Trojan

A *Trojan* is a malicious program disguised as a normal-looking program. While a virus attaches itself to another program, a Trojan is embedded within the program itself. When the normal-looking program runs, the embedded code runs also and causes damage. Trojans do not replicate themselves the way viruses and worms do. Trojans often spread through emails or worms.

Drive-by Download

A drive-by download can be one of two things [1]:

- A download that you authorized but without understanding the consequences.
- A download that happens without your consent or even your knowledge.

A drive-by download involves any type of downloaded file containing a virus or Trojan. It often occurs when you visit a website, read an e-mail message, or click on a deceptive pop-up window.

Today's malware is more complex than ever. Understanding the various types of malware is just the first step in protecting your system from cyber threats. Make sure you have security software installed on your system, and don't click on suspicious links!

出 处

1. Drive-by download, Wikipedia, http://goo.gl/tNkB

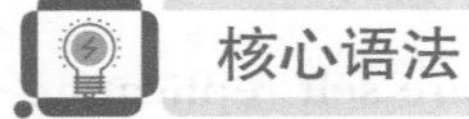

核心语法

熟悉 # 关系代名词。# 关系代名词可以连接两个句子，兼具 # 连词和 # 代名词的作用。例文将介绍多种关系代名词。# 关系代名词皆有章可循，理解原理并结合例句学习将事半功倍。

- All of these are **malicious programs** (**that** can harm our computers). 这些都是恶意程序/会破坏电脑的。
- Worms are **malware** (**that** run by themselves). 蠕虫是恶意软件/自己运行的。
- **They are self-replicating**, **which** means they do not need host programs. 它们自我复制，这意味着/它们不需要主机程序。
- It is a **malicious program** ([**that** is] disguised as a normal-looking program). 它是恶意程序/被伪装为正常程序的。
- It involves any type of **downloaded file** ([**which** is] containing a virus). 它包含所有下载文件/包括病毒的。

熟悉 # 所有格关系代名词。所有格关系代名词较难分析，建议将两个句子拆开理解。

- **Any program** is malware. + **Its** intention is malicious. 所有程序都是恶意软件。+ 它们的意图都是恶意的。

  = **Any program** (**whose** intent is malicious) is malware. =所有程序/它们的意图是恶意的/是恶意软件。

熟悉 # 关系副词。# 关系副词可以连接两个句子，兼具 # 连词和 # 副词的作用。原理与 # 关系代名词类似，但 # 先行词和 # 关系副词的省略条件略有不同。以下例句适用于两者。

● Trojans do not replicate themselves **the way**. + Viruses replicate themselves **in the way**. “特洛伊木马”不自我复制/用这种方式。+ 病毒自我复制/用这种方式。

= Trojans do not replicate themselves **the way**. + Viruses do **in the way**. = “特洛伊木马”不自我复制/用这种方式。+ 病毒/用这种方式。

= Trojans do not replicate themselves **the way** viruses do. = “特洛伊木马”不自我复制/用这种方式/病毒用的。

= Trojans do not replicate themselves **how** viruses do.

▶ The way和how不能同时使用。二者只能选一。

熟悉 # 分词的惯用型或者 # 独立分词从句。惯用型需要熟记。例文中介绍的惯用型使用频率较高，常被视为 # 前置词。

● **Depending on** how it infects a system, malware fits into one of the following categories. 根据其对系统的感染程度，恶意软件被划分属以下范围。

● Malware is classified into several types **based on** several criteria. 恶意软件被分为很多类/根据众多标准。

● Any hostile programs, **including** viruses and worms, are all malware. 所有敌对程序，包括病毒和蠕虫，都是恶意软件。

单词&短语

consider A to V 将 A 视为 V
malicious 恶意的
harm 伤害
behave 行动
combination 组合
term 术语，单词
suggest 建议，暗示
intent 意图
hostile 敌对的
intrusive 侵入
including ~ 包括 ~
be classified into ~ 被归类为 ~
based on ~ 基于 ~
criteria（复数）标准
criterion（单数）标准
infection method 感染方式
generally 一般地
fit into 适合 ~
attach A to B 让 A 附属于 B
reproduce 复制
additional 附加的
spread 传播
email attachment 附件
by oneself 独立的

self-replicating 自我复制
self-propagating 自我传播
travel through ~ 通过 ~ 旅行
on one's own 独立的，自主的
replicate 复制
interaction 相互作用，交互
be disguised as ~ 被伪装为 ~
embed 内置
normal-looking 正常显示
embedded code 内置代码
authorize 许可，授权
consequence 结果
consent 同意
knowledge 知识
involve 包含
contain 包含
occur 发生
deceptive 虚伪的，骗人的

技术术语

spyware 间谍软件意在秘密收集个人或组织信息。未经个人或组织允许的情况下，将计算机资源传送给第三方。间谍软件是恶意软件的一种，它采取众多方式掩盖真身，很难发现。（为了获取密码或者信用卡信息）拦截用户键盘的 Key Logger 就是典型的间谍软件。

根据提示完成句子

Many PC users consider /
很多PC用户认为/

viruses, malware, spyware, adware, worms, and Trojans /
病毒、恶意软件、间谍软件、广告软件、蠕虫和“特洛伊木马”/

___ __ __ __ ____ _____. / While all of these are ________ programs /
都一样。/ 尽管它们都是恶意程序/

that can ____ our computers, / ____ _______ / ___________. /
能够破坏我们的电脑，/ 但各自表现/ 不同。/

Malware is a ___________ of two words: / '_________' and '________'. /
恶意软件是两个单词的组合：/ “恶意”和“软件”。/

As the ____ _________, / any program / _____ ______ is _________ /
顾名思义，/ 所有程序/ 它的意图是恶意的/

is malware. / ____ __, / any _______ or _________, /

就是恶意软件。/ 所有的敌对性或者侵入性程序，/

including viruses, spyware, adware, worms, and Trojans, /

包括病毒、间谍软件、广告软件、蠕虫、“特洛伊木马”，/

are all malware. /

都是恶意软件。

思考题

解答题

1 （理解）以下哪种恶意软件可以自我复制？

ⓐ 病毒　　ⓑ“特洛伊木马”

ⓒ 广告软件　　ⓓ 间谍软件

2 （论述）请列举恶意软件的类型。

讨论

1 Android手机用户因为强迫下载（drive-by download，DBD）攻击深受其害，请列举解决办法。

课堂总结

1 学习#关系代名词的概念和例句。

2 学习#所有格关系代名词的概念和例句。

3 熟记#分词的惯用型或#独立分词从句。

答案

1 a. 2 病毒、蠕虫、“特洛伊木马”、广告软件、用户下载

翻译

Viruses and Malware: What's the Difference?
病毒与恶意软件区别何在?

Many PC users consider viruses, malware, spyware, adware, worms, and Trojans /
很多 PC 用户认为病毒、恶意软件、间谍软件、广告软件、蠕虫、和“特洛伊木马”/
all to be the same thing. / While all of these are malicious programs / that can harm our computers, /
都一样。/ 尽管它们都是恶意程序 / 能够破坏我们的电脑，/
each behaves / differently. /
但各自表现 / 不同。/

Malware is a combination of two words: / 'malicious' and 'software'. /
恶意软件是两个单词的组合：/“恶意”和“软件”。/
As the term suggests, / any program / whose intent is malicious / is malware. / That is, /
顾名思义，/ 所有程序 / 它的意图是恶意的 / 就是恶意软件。/ 换言之，/
any hostile or intrusive programs, / including viruses, spyware, adware, worms, and Trojans, /
所有敌对性或者侵入性程序，/ 包括病毒、间谍软件、广告软件、蠕虫、“特洛伊木马”，/
are all malware. /
都是恶意软件。/

Malware is classified into several types / based on several criteria. /
恶意软件可以划分为很多类 / 根据不同标准。/
The most commonly used criterion is infection method. /
最常用的分类标准是感染方式。/
Depending on how it infects a system, / malware generally fits into one of the following categories: /
根据感染系统的方式，/ 恶意软件一般可以分为如下几种。/

Virus / 病毒

A virus is malware / that attaches itself / to (or infects) other programs or files. /
病毒是恶意软件 / 依附在 /（或者感染）其他程序或文件。/
When a user runs an infected program (or a host program), / the virus runs also /
用户运行受感染的程序时（或者主机应用程序），/ 病毒也开始运行 /
and performs malicious activities. / It also reproduces itself / by attaching itself /
并开始进行恶意活动。/ 它也进行自我复制 / 通过依附在 /
to additional files. / Viruses spread through human actions / such as forwarding an infected file, /
附加文件上。/ 病毒通过人的行为传播 / 例如转发被病毒感染的文件，/
opening an infected email attachment, /
打开被感染的邮件附件，/
or downloading and running a malicious program / from the Internet. /
或者下载并运行恶意程序 / 从网上。/

Worms / 蠕虫

Worms are malware / that run / by themselves. / They are self-replicating and self-propagating, /
蠕虫是恶意软件 / 它运转 / 自行。/ 它们自我复制并且自我传播，/
which means / they do not need host programs / or human actions / to spread. /
这意味着 / 它们不需要主机应用程序 / 或者借助人类行为 / 传播。/
Worms travel through networks / on their own / while replicating themselves. /
蠕虫借助网络传播 / 自主地 / 同时进行自我复制。/
They can spread very fast / because there is no need / for human interaction. /
它们能迅速传播 / 因为不需要 / 与人类交互。/

Trojan / "特洛伊木马"

A Trojan is a malicious program / disguised as a normal-looking program. /
"特洛伊木马"是恶意程序 / 伪装为正常程序。/
While a virus attaches itself / to another program, / a Trojan is embedded /
病毒需要自我依附于 / 其他程序，/ 但"特洛伊木马"内置于 /
within the program itself. / When the normal-looking program runs, /
程序内部。/ 一旦这个看似正常的程序运行，/
the embedded code runs also / and causes damage. / Trojans do not replicate themselves /
内置代码也一同运行 / 并且开始破坏。/"特洛伊木马"不进行自我复制 /
the way viruses and worms do. / Trojans often spread / through emails or worms. /
以病毒和蠕虫的方式。/"特洛伊木马"经常传播 / 通过邮件或者蠕虫。/

Drive-by Download / DBD攻击

A drive-by download can be one of two things [1]: /
DBD 攻击有两种触发机制：/
A download that you authorized / but without understanding the consequences. /
一种是用户授权下载 / 但是在不知道结果的状况下。/
A download that happens / without your consent / or even your knowledge. /
一种是强制下载 / 未经用户同意 / 或者在用户不知情的状况下。/
A drive-by download involves any type of downloaded file / containing a virus or Trojan. /
DBD 攻击涉及所有格式的下载文件 / 包括病毒或者"特洛伊木马"。/
It often occurs / when you visit a website, / read an e-mail message, /
它经常出现 / 当你浏览网站 / 阅读邮件，/
or click on a deceptive pop-up window. / Today's malware is more complex / than ever. /
或者点击虚假弹窗时。/ 当今的恶意软件更加复杂 / 相比之前。/
Understanding the various types of malware / is just the first step / in protecting your system /
理解各种类型恶意软件 / 仅是第一步 / 对于保护你的系统 /
from cyber threats. / Make sure / you have security software / installed on your system, /
免受网络攻击。/ 确保 / 你安装了安全软件 / 在系统中，/
and don't click on suspicious links! /
切勿点击可疑链接！/

预装的众多计算机程序
Adware, Spyware, Ransomware, and What Else?

Malware is short for "malicious software."[1] It is designed to damage or disrupt a system. People often call any dangerous software a virus, but there are many different types of malware. Viruses are just one form of malware.

In fact, various criteria are used to classify malware. One of the most common criteria is the infection (or spreading) method of the malware. For example, viruses, worms, Trojans, and drive-by downloads spread in different ways.

Another commonly used criterion is the behavior of the malware. Once malware is on your system, it can do many things. Some behaviors do little harm, but others can be nasty. The following behavior-based malware types are common:

Adware

Adware displays unwanted advertisements. Adware is often bundled with downloaded free software. When you install the software, adware is installed along with it. Usually, you "agree" to install the adware during the software installation process. So pay attention to the checkbox during the installation process. Generally adware can be removed if you uninstall the software.

Spyware

Spyware monitors your activities, steals your information and sends it to another party without your knowledge. Most spyware is designed to track your Web browsing history and online purchasing habits, but some malicious programs can steal your identity and corrupt your data. For example, keyloggers record all keystrokes to steal your passwords or credit card numbers. Fortunately, most spyware can be removed by anti-spyware tools.

Ransomware

Ransomware locks up a computer and demands ransom. It holds your computer hostage until the ransom is paid. It's a very popular way to make money for cyber criminals. Security experts advise never to pay the ransom for three reasons. First, there's no guarantee of getting your files back even after you pay. Second, criminals may load additional viruses on your system. Third, you should never reward bad behaviors. Instead, experts recommend anti-malware software and frequent backups.

Scareware

Scareware tricks you into buying or installing potentially dangerous software. A common trick is to display a message that says, "Your computer is infected with a virus, so download this antivirus program to remove it." Of course, it's a fake antivirus program!

Nowadays malware is getting more and more clever. Often it does more than one thing to damage your system. For example, adware may monitor your online activities or open a backdoor to invite other malware. So be extra careful at all times. Install antivirus software and back up your files often. Remember, prevention is better than cure!

出 处

1. Malware, Wikipedia, http://goo.gl/Gl9tv

核心语法

熟悉 # 句式 5〈主语 + 动词 + 宾语 + 宾补〉。本节将使用 # 宾语和 # 宾补作同位语的句子。该结构类似 # 句式 4〈主语 + 动词 + 间接宾补 + 直接宾语〉。从最后两个例句可以看出，本句式有多种解释，根据上下文理解即可。

- People call any dangerous software a virus. 人们称/所有的危险软件为病毒。
- Ransomware holds your computer hostage. 勒索软件/用你的电脑要挟。
- Mom made **me a doll**. 妈妈给我做了个玩具。（给~ 做~）
- The witch made **me a doll**. 魔女将我变为玩具。（我 = 玩具）

熟悉 # 不定式 to 作 # 名词、# 形容词、# 副词的句子。从最后一个例句可以看出，它有多种解释。根据上下文理解即可。

- A common trick is **to display** a message. 最常见的骗局就是展示信息。
- You agree **to install** the adware. 你同意/安装广告软件。
- Security experts advise **never to pay** the ransom. 安全专家建议/千万不要付赎金。
- It's a very popular way **to make** money. 这是一个很受欢迎的方式/赚钱的。
- Malware is designed **to damage** a system. 恶意软件诞生的目的是/摧毁系统。
- It record all keystrokes **to steal** your passwords. 它记录所有键盘输入/以窃取你的密码。
- Adware opens a backdoor **to invite** other malware. 广告软件打开后门/给其他恶意软件。或者广告软件打开后门/让其他恶意软件进入。

熟悉 # 被动语态〈be+ 过去分词〉句型。如果理解了主动语态，就不需要死记硬背 be designed to 这类惯用表达。

- Malware **is designed** to disrupt a system (by someone). 恶意软件被设计的目的是/

摧毁系统。

= Someone **designs** malware to disrupt a system. =有人设计恶意软件/为了摧毁系统。

- Various criteria **are used** to classify malware (by people). 众多标准被使用/为了将恶意软件进行分类。

  = People **use** various criteria to classify malware. =人们使用众多标准/将恶意软件分类。

- Adware **is bundled** with free software (by people). 广告软件被绑定/和免费软件一起。

  = People **bundle** adware with free software. =人们将广告软件绑定/和免费软件。

单词&短语

be short for ~ 简称为 ~，简写为 ~
malicious software 恶意软件
call A B 将 A 称为 B
be designed to V 被设计的目的 ~
damage 损害
disrupt 扰乱
various 各种各样的
criteria 标准（复数）
criterion 标准（单数）
classify 分类
infection method 感染方式
spreading method 扩散方式
do harm 伤害
nasty 难对付的
unwanted 不希望
be bundled with 被捆绑
along with ~ 和 ~ 一起，跟随 ~
installation process 安装过程
pay attention to ~ 关注 ~
generally 一般地
activity 活动
steal 窃取
track 追踪
identity 身份
corrupt 破坏
lock up 锁
ransom 赎金
hold a person hostage 将 ~ 作为人质

技术术语

malware vs. virus 恶意软件与病毒的前者指网络中泛滥的恶意代码，它感染计算机（或者手机）后，大量发送垃圾信息、窃取银行密码等，令人生厌；后者是通过感染文件传播病毒的代码。真正的计算机病毒很少见，多是为了实施犯罪行为，计算机病毒只是中间媒介。1980~1990 年，PC 一度盛行，那时因为病毒是恶意软件的主要形态，所以人们倾向于将恶意程序称为“病毒”，这种习惯一直保持到现在。

根据提示完成句子

Malware is ______ ___ “malicious software.”[1] / It is designed /
Malware 是malicious software的简称。/ 它最初的设计目的是/

to ______ or _______ a system. / People often call /
摧毁或扰乱系统。/ 人们通常/

any __________ _______ a _____, /
将所有危险软件称为“病毒”，/

but there are many __________ _____ of malware. /
但恶意软件有很多类型/

Viruses are just ___ ____ of malware. /
病毒仅是恶意软件中的一种。/

________ / malware is _______ ____ ___ ____ ______. /
最近/ 恶意软件变得越来越聪明。/

Often it does ____ ____ ___ _____ / to ______ ____ ______. /
它经常做很多事情/ 以摧毁系统。/

For example, / adware may ______ ___ ______ __________ /
例如，/ 广告软件可能监测用户的线上活动/

or ____ _ ________ / to ______ _____ _______. /
或打开后门/ 以供其他恶意软件进入。/

So __ _____ _______ / __ ___ ______. /
因此需要格外注意/ 随时。/

Install antivirus software / and back up your files / often. /
安装杀毒软件/ 并备份文件/ 经常。/

Remember, / __________ is ______ ____ ____! /
切记，/ 防患于未然!

思考题

解答题

[1] （理解）以下哪种恶意软件能带来直接经济损失？

ⓐ 广告软件　ⓑ 间谍软件

ⓒ 诈骗软件　ⓓ 虚假杀毒软件

[2] （理解）以下哪种恶意软件带来的物质损失最低？

ⓐ 广告软件　ⓑ 间谍软件

ⓒ 诈骗软件　ⓓ 虚假杀毒软件

[3] （论述）为阻止恶意软件入侵，个人应该采取哪些应对措施？

[4] （论述）为避免被广告软件欺骗，安装阶段有哪些注意事项？

讨论

[1] 如果在工作中遇到诈骗软件传播导致公司网站出现严重问题，应当由谁负责？该如何应对？

[2] 人们最近更关注恶意软件而非病毒的原因何在？

课堂总结

[1] 学习#句式5和例句。复习#句式4。

[2] 学习#不定式和#动名词作#宾语的#动词用法和例句。

[3] 学习#被动态和#主动态。

答案

[1] **b**.（窃取密码和信用卡信息的可能性很大） [2] **a**.（多数情况下只展示广告） [3] 安装杀毒软件，安装最新版本的操作系统和浏览器/Flash/Java插件，定期备份，禁止安装从网上下载的软件。 [4] 在安装应用的过程中，要认真阅读附加应用安装许可条款，要了解捆绑安装的软件类型。

翻译

Adware, Spyware, Ransomware, and What Else?
预装的众多计算机程序

Malware is short for "malicious software."[1] / It is designed / to damage or disrupt a system. /
Malware 是 malicious software 的简称。/ 它的设计目的是 / 摧毁或扰乱系统。/
People often call / any dangerous software a virus, / but there are many different types of malware. /
人们通常将 / 所有危险软件称为 "病毒" / 但恶意软件有很多种类。/
Viruses are just one form of malware. /
病毒仅是恶意软件中的一种。/

In fact, / various criteria are used / to classify malware. /
事实上，/ 人们用各种标准 / 区分恶意软件。/
One of the most common criteria is the infection (or spreading) method of the malware.
最常见的标准是感染（或者传播）方式。/
For example, viruses, worms, Trojans, and drive-by downloads spread / in different ways. /
例如，病毒、蠕虫、"特洛伊木马" DBD 攻击传播 / 以不同方式。/

Another commonly used criterion / is the behavior of the malware. / Once malware is on your system, /
另一种常见标准是 / 恶意软件的行为。/ 一旦恶意软件在你的系统里，/
it can do many things. / Some behaviors / do little harm, but others can be nasty.
它就可以做许多事情。/ 一些行为是无害的 / 但其他却很难对付。/
The following behavior-based malware types are common:
以下是常见的恶意软件类型。/

Adware / 广告软件

Adware displays unwanted advertisements. / Adware is often bundled /
广告软件展示人们不愿意看到的广告。/ 广告软件经常 /
with downloaded free software. / When you install the software, /
和免费下载软件捆绑下载。/ 当你安装软件时，/
adware is installed / along with it. / Usually, / you "agree" /
广告软件就被安装 / 和它一起。/ 通常，/ 你 "同意" /
to install the adware / during the software installation process. /
安装广告软件 / 在下载软件的过程中。/
So pay attention / to the checkbox / during the installation process. /
因此要关注 / 勾选框 / 在安装过程中。/
Generally / adware can be removed / if you uninstall the software. /
一般来说 / 广告软件可以删除 / 如果你卸载软件。/

Spyware / 间谍软件

Spyware monitors your activities, / steals your information / and sends it to another party /
间谍软件监视你的举动，/ 窃取你的信息 / 并将其发送给其他人 /

without your knowledge. / Most spyware is designed / to track your Web browsing history /
在你不知情的时候。/ 多数间谍软件的设计目的是 / 追踪你的网页浏览记录 /
and online purchasing habits, / but some malicious programs can steal your identity /
和网上购物习惯，/ 但有些恶意程序可以窃取你的身份信息 /
and corrupt your data. / For example, / keyloggers record all keystrokes /
并且破坏你的数据。/ 例如，/Key Logger 记录所有键盘输入 /
to steal your passwords / or credit card numbers. / Fortunately, /
以窃取你的密码 / 或者信用卡号。/ 幸运的是，/
most spyware can be removed / by anti-spyware tools. /
多数间谍软件可删除 / 用杀毒工具。/

Ransomware / 诈骗软件

Ransomware locks up a computer / and demands ransom. / It holds your computer hostage /
诈骗软件锁住电脑 / 并索要赎金。/ 它将你的电脑当作“人质”/
until the ransom is paid. / It's a very popular way / to make money for cyber criminals. /
直到支付赎金。/ 这是一种流行方式 / 网络罪犯赚钱的。/
Security experts advise / never to pay the ransom / for three reasons. /
安全专家建议 / 不要支付赎金 / 有 3 种原因。/
First, / there's no guarantee / of getting your files back / even after you pay. /
第一，/ 不能保证 / 拿回你的文件 / 即使支付了赎金。/
Second, / criminals may load additional viruses / on your system. /
第二，/ 罪犯可能又附加安装了病毒 / 在你的系统。/
Third, / you should never reward bad behaviors. / Instead, /
第三，/ 绝不可以姑息恶意行为。/ 正确的做法是，/
experts recommend anti-malware software / and frequent backups. /
专家推荐反恶意软件 / 和定期备份。/

Scareware / 虚假杀毒软件

Scareware tricks you / into buying / or installing potentially dangerous software. /
虚假杀毒软件欺骗你 / 购买 / 或者安装具有潜在危险的软件。/
A common trick is / to display a message / that says, /
常见的伎俩是 / 输出信息 / 称：/
"Your computer is infected / with a virus, / so download this antivirus program /
“你的计算机感染了 / 病毒，/ 所以要下载这个杀毒程序 /
to remove it." / Of course, / it's a fake antivirus program! /
以清除病毒。”/ 当然，/ 这是假的杀毒程序！ /

Nowadays / malware is getting more and more clever. / Often it does more than one thing /
最近 / 恶意软件越来越聪明。/ 它经常做很多事情 /
to damage your system. / For example, / adware may monitor your online activities /
以摧毁系统。/ 例如 / 广告软件可能会监视你的线上活动 /
or open a backdoor / to invite other malware. / So be extra careful / at all times. /
或者打开后门 / 以供其他恶意软件进入。/ 需要格外注意 / 随时。/
Install antivirus software / and back up your files / often. / Remember, / prevention is better than cure! /
安装杀毒软件 / 并备份文件 / 经常。/ 切记，/ 防患于未然！ /

第二部分

无人机/机器人

从无人驾驶汽车到机器人记者，计算机取代人的趋势越来越明显。这部分主要讲述无人机和机器人的主要应用领域，以及IT高手对此事的看法。

谷歌与Facebook 的空中争霸战

Internet Space Race: Google vs. Facebook

While the military has been using drones in its various wars and conflicts around the world since the 1970s, drones were mostly unknown to the public until Amazon CEO Jeff Bezos revealed a secret R&D project on a CBS TV newsmagazine program, *60 Minutes*, in December 2013 [1]. Amazon's *Octocopter* drones will fly packages directly to your doorstep in 30 minutes within five years, said Jeff Bezos. The Internet was lit up with discussion, speculation, and concerns. Clearly it was a kickstart for the drone industry. Since then, companies like DHL, Coke, Domino's Pizza, and Alibaba have jumped on the drone-delivery bandwagon.

While Amazon and these companies are focusing on delivery drones—and struggling to get approval by the Federal Aviation Administration (FAA) — tech giants Google and Facebook are looking to drone technology to bri ng Internet access to everybody and connect the whole world. Their mot ivations are not entirely altruistic. More than 60% of the world's populati on does not have Internet access [2], which translates to four billion pote ntial new users for Google and Facebook. These two companies have alr eady taken their battle for Internet dominance to the sky.

Google Project Loon & Project Titan

Project Loon has been making great strides since it was launched in June 2013. "Project Loon is a network of balloons traveling on the edge of space, designed to connect people in rural and remote areas, help fill coverage gaps, and bring people back online after disasters," says the project's website [3]. The Internet-beaming balloons, which lasted only a few days during the first New Zealand tests, now can stay afloat for six months at a time and deliver LTE data speed to devices on the ground.

Project Titan, which complements Project Loon, is also making progress since Google acquired Titan Aerospace, a drone manufacturer that makes high-altitude drones, in early 2014. Titan's solar-powered drones will stay in the upper atmosphere for five years at a time and beam down Internet access to remote areas around the globe.

Facebook Connectivity Lab & Internet.org

In August 2013, Facebook CEO Mark Zuckerberg announced Internet.org, a partnership between Facebook and seven mobile phone companies, with an aim to make the Internet accessible and affordable to all. Then in March 2014, Zuckerberg revealed on his blog that Facebook's Connectivity Lab is working to "build drones, satellites, and lasers to deliver the Internet to everyone."[4]

Yael Maguire, engineering director at Connectivity Lab, said the drones will be "roughly the size of a commercial aircraft, like a 747" and will "have to fly for months, or perhaps years at a time."[5] Facebook plans to start testing its drones in 2015.

出 处

1. Amazon Unveils Futuristic Plan Delivery by Drone, 60 Minutes, http://goo.gl/vMF31g
2. Offline and falling behind: Barriers to Internet adoption, http://goo.gl/lljSlx
3. Project Loon, http://www.google.com/loon
4. Mark Zuckerberg's Facebook Posting, https://goo.gl/9wO02t
5. Facebook Further Reveals Plans for Internet-Connected Drones, http://goo.gl/rNndsX

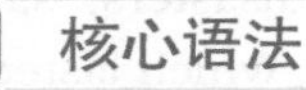

核心语法

熟悉 # 现在完成时态〈have/has+ 过去分词〉。指过去发生的事情对现在造成的影响。

- The military **has been using** drones **since** the 1970s. 军队一直在使用无人机/从20世纪70年代起。（=现在也在使用）
- Companies **have jumped** on the drone-delivery bandwagon. 公司已经赶上了无人机送货的潮流。（=所以现在是用无人机送货的状态）
- These two companies **have already taken** their battle to the sky. 这两家公司已经展开竞争/在空中。（=所以现在已经展开了空中战）

熟悉 # 关系代名词 which。# 先行词可以是物或者句子。# 关系代名词有规律可循，理解原理并结合例句学习将事半功倍。

- More than 60% of the world's population does not have Internet access, **which** translates to four billion potential new users. 全球60%以上的人无法上网，这（=代指前句）意味着将有 40亿潜在新用户。
- **Project Titan**, **which** complements Project Loon, is also making progress. 太阳能无人机项目——它是高空热气球项目的补充——也取得了进展。
- Project Loon is a network of **balloons** ([**which** are] traveling on the edge of space). 高空热气球项目是气球网络/它在太空边缘运转。
- Project Loon is **a network** of balloons, [**which** is] designed to connect people. 高空热气球项目是气球网络，它的设计目的是为了人与人之间的联系。

长句阅读提示。

- **While** the military has been using drones in its various wars and conflicts around the world since the 1970s, / drones were mostly unknown to the public / **until** Amazon CEO Jeff Bezos revealed a secret R&D project on a CBS TV newsmagazine program, *60 Minutes*, in December 2013.

▶ 句子变长的原因之一是 # 连词的存在。根据连词可将上述句子拆分为 3 个短句进行理解。〈前置词 + 名词〉构成 # 前置句较容易混淆视听，建议先将其去掉，简化句子结构。

● Project Titan, which complements Project Loon, is also making progress / **since** Google acquired Titan Aerospace, a drone manufacturer that makes high-altitude drones, in early 2014.

▶ 逗号前后有#插入句，所以整个句子看起来冗长且复杂。此时建议先去掉插入句。上文去掉#插入句和#前置句后，可分为两个短句。

单词&短语

various 各种各样的
conflict 冲突，纷争
mostly 主要、一般
the public 普通公民、大众
reveal 揭露、显示
lit up with 点燃
discussion 讨论
speculation 推测
concern 担心
clearly 明确的
kickstart 启动
jump on the bandwagon 赶时髦
focus on 集中于
struggle to V 挣扎做 ~
approval 认可
aviation 航空
look to ~ 考虑
altruistic 利他的
translate to ~ 意味着 ~
take A to B 把 A 带往 B
dominance 支配、占主导地位
make great strides 有长足的进步
rural area 农村
remote area 偏远地区
coverage gap 覆盖缺口
~beaming 轰炸 ~
last 持续
afloat 悬浮
complement 补充
make progress 进展
acquire 取得、获得、收购
manufacturer 制造商
altitude 高度、海拔
solar-powered 太阳能供电的
announce 发布
partnership 合作关系
with an aim to V 目的是 ~
accessible 可接近的、可使用的
affordable 价格实惠的

根据提示完成句子

While Amazon and these companies are ________ /

当亚马逊和这些企业关注/

__ _________ ______ / – and struggling to ___ ________ /

无人机送货/ 并且在努力得到认可时/

by the Federal Aviation Administration (FAA) - /

美联邦航空局的/

tech giants Google and Facebook are looking to drone technology /

技术巨匠谷歌和Facebook却在关注无人机技术/

to _____ ________ ______ / to everybody /

以提供网络/ 给所有人/

and _______ ___ _____ _____. /

和连接全球。/

Their motivations are not ________ _________. /

他们的动机并非没有私心。/

More than 60% of the world's ________ does not have ______ ____ [2], /

全球60%以上的人口无法上网，/

which _________ __ four billion ________ ___ ____ /

这意味着有40亿潜在新用户/

for Google and Facebook. /

给谷歌和Facebook。/

These two companies have already _____ _____ ______ /

这两家企业已经开始了竞争/

for _______ _________ / __ ___ ___. /

为了抢占网络先机/ 在空中。/

思考题

解答题

1 （理解）Facebook普及网络时不会使用以下哪一项技术？

ⓐ 无人机　　ⓑ 人工卫星

ⓒ 激光　　ⓓ 气球

2 （论述）比较谷歌的太阳能无人机项目和高空热气球项目。

讨论

1 谷歌和Facebook在网络服务方面展开了激烈的空间战，目的何在？这与谷歌最近开始实施的仅用20美元就可以无限通话的“虚拟运营商服务有何关联”？

2 谷歌和Facebook的空间战对超高速网络尚不完备的中国有何影响？

课堂总结

1 学习#现在完成时态和#现在完成进行时态的概念和例句。

2 学习#关系代名词的概念和例句。

3 学习#准关系代名词的概念和例句。

答案

1 d.（谷歌的高空热气球项目使用气球） 2 高空热气球项目使用位于太空边缘的气球运行无线网络，无人机使用在高纬度运行的无人机运行无线网络。

翻译

Internet Space Race: Google vs. Facebook
谷歌与Facebook 的空中争霸战

While the military has been using drones / in its various wars / and conflicts / around the world /
尽管军队一直使用无人机 / 在各种战争 / 和冲突中 / 全球范围的 /
since the 1970s, / drones were mostly unknown / to the public /
从 20 世纪 70 年代起，/ 但无人机广为人知 /
until Amazon CEO Jeff Bezos revealed a secret R&D project /
要一直等到亚马逊首席执行官杰夫·贝佐斯揭露神秘的 R&D 项目 /
on a CBS TV newsmagazine program, / 60 Minutes, / in December 2013 [1]. /
在 CBS TV 时事周刊节目 /60 分，/ 在 2013 年 12 月。/
Amazon's Octocopter drones will fly packages / directly / to your doorstep / in 30 minutes /
亚马逊的 Octocopter 无人机能装载包裹 / 径直 / 飞到门前 / 在 30 分钟内 /
within five years, / said Jeff Bezos. / The Internet was lit up / with discussion, speculation, and concerns. /
在 5 年内，/ 杰夫·贝佐斯指出。/ 网络充斥着 / 争论、猜测、担忧。/
Clearly it was a kickstart / for the drone industry. /
很明显，这是一个开始 / 无人机产业的。/
Since then, / companies / like DHL, Coke, Domino's Pizza, and Alibaba /
从那时起，/ 企业 / 诸如 DHL、可口可乐、达美乐披萨、阿里巴巴等 /
have jumped on the drone-delivery bandwagon. /
已经开始进入无人机送货领域。/

While Amazon and these companies are focusing / on delivery drones /
当亚马逊和这些企业关注 / 无人机送货 /
- and struggling to get approval / by the Federal Aviation Administration (FAA) - /
并在努力得到认可时 / 美联邦航空局的 /
tech giants Google and Facebook are looking to drone technology /
技术巨匠谷歌和 Facebook 却在关注无人机技术 /
to bring Internet access / to everybody /
以提供网络 / 给所有人 /
and connect the whole world [POCKET]. / Their motivations are not entirely altruistic. /
并连接全球。/ 他们的动机并非没有私心。/
More than 60% of the world's population does not have Internet access [2], /
全球 60% 以上的人口无法上网，/
which translates to four billion potential new users / for Google and Facebook. /
这意味着有 40 亿潜在新用户 / 给谷歌和 Facebook。/
These two companies have already taken their battle / for Internet dominance / to the sky. /
这两家企业已经开始了竞争 / 为了抢占网络先机 / 在空中。/

Google Project Loon & Project Titan / 谷歌高空热气球项目和无人机项目

Project Loon has been making great strides / since it was launched / in June 2013. /
高空热气球项目已取得了长足的进展 / 从它开始启动 / 在 2013 年 6 月。/
“Project Loon is a network of balloons / traveling on the edge of space, /
“高空热气球项目是气球网络 / 在太空边缘运行，/
designed / to connect people / in rural and remote areas, / help fill coverage gaps, /
设计目的在于 / 人际沟通 / 位于农村和偏远地区的，/ 帮助缩小差距，/
and bring people back online / after disasters,” / says the project’s website [3]. /
以及带他们重返网络 / 在灾难过后。”/ 出自项目官网。/

The Internet-beaming balloons, / which lasted / only a few days / during the first New Zealand tests, /
这些网络气球，/ 持续 / 仅有几天 / 在首次新西兰测试期间，/
now can stay afloat / for six months / at a time / and deliver LTE data speed / to devices / on the ground. /
现在可以悬浮 /6 个月 / 一次 / 并且传输 LTE 数据速度 / 到设备 / 地面的。//

Project Titan, / which complements Project Loon, / is also making progress /
无人机项目，/ 是高空热气球项目的补充，/ 也取得了进展 /
since Google acquired Titan Aerospace, / a drone manufacturer / that makes high-altitude drones, /
从谷歌收购泰坦航空以后，/ 无人机制造商 / 制造高纬度无人机，/
in early 2014. / Titan’s solar-powered drones will stay / in the upper atmosphere / for five years /
在 2014 年初。/ 泰坦航空的太阳能供电无人机可停留在 / 高层大气 /5 年 /
at a time / and beam down Internet access / to remote areas around the globe. /
一次 / 并且将网络传输给 / 地球上的偏远地区。/

Facebook Connectivity Lab & Internet.org / Facebook 互联实验室和Internet.org

In August 2013, / Facebook CEO Mark Zuckerberg announced Internet.org, /
2013 年 8 月，/Facebook 首席执行官马克 · 扎克伯格公布了 Internet.org，/
a partnership between Facebook and seven mobile phone companies, / with an aim /
Facebook 和 7 大手机公司的合作关系，/ 旨在 /
to make the Internet accessible and affordable / to all. / Then in March 2014, /
能接触到网络且能承担起费用 / 让所有人。/ 在 2014 年 3 月，/
Zuckerberg revealed on his blog / that Facebook’s Connectivity Lab is working /
扎克伯格在他的博客上称 /Facebook 互联实验室的目的 /
to “build drones, satellites, and lasers / to deliver the Internet to everyone.”[4] /
在于打造 “无人机、人工卫星、激光 / 以让所有人都可以上网。”/

Yael Maguire, / engineering director at Connectivity Lab, /
耶尔 · 马奎尔，/ 互联实验室网络工程总监，/
said the drones will be “roughly the size of a commercial aircraft, like a 747” /
说无人机 “大小类似 747 商用客机”/
and will “have to fly / for months, / or perhaps years at a time.”[5] /
并且 “可以一次飞行，/ 数月 / 甚至数年。”/
Facebook plans / to start testing its drones in 2015. /
Facebook 计划 / 在 2015 年启动测试这种无人机。/

无人机的五种特色用途
Five Cool Uses for Drones

When people talk about drones, Amazon's delivery drones often lead the discussion. However, drones, or unmanned aerial vehicles (UAVs), were used long before Jeff Bezos announced his plan. Originally drones were used for military and intelligence applications like spying and bombing. Today more and more civilian drones are being used in surprising areas.

Here are five cool uses for drones:

Hurricane Hunting

Drones can fly into the heart of a storm and spy on it as the storm develops. A team at the University of Florida is using a swarm of six-inch-long drones to track tropical storms [1]. The drones are launched by a laptop, use minimal power, and can be carried by wind or water current. They ride through a massive hurricane and collect data such as temperature, pressure, humidity, and location. Their reports could help scientists improve their storm-prediction powers.

3D Mapping

Small, lightweight drones fly over landscapes and take thousands of digital images. Then software pieces them into 3D maps. A 3D model of the Ambrym Volcano on the Pacific island of Vanuatu is a fascinating example [2]. The volcano was deeper than the Empire State Building and full of hot lava! Geologists used a drone and special drone mapping software to render the 3D model. The model will help scientists understand volcanoes and their risks.

Protecting Wildlife

The U.S. government already uses drones to protect its lands and wildlife. Several agencies are using the military Raven drones to monitor wildlife populations or map roads and wetlands for land management purposes. An Orangutan Conservancy also uses drones to identify the animals' distribution and density in Indonesia and Malaysia [3].

Farming

Farmers are using drones to monitor fields, increase yields, and save money. Drones with digital cameras help farmers identify problem areas for closer inspection, but now drones with infrared light cameras can inspect crop health by detecting photosynthetic efficiency in various plants [4].

Search and Rescue

First responders are also getting help from drones. Search-and-rescue drones can fly over dangerous areas and search for survivors. Lifeguard drones reach swimmers faster than human lifeguards. Last year a prototype of an *ambulance drone* was introduced by a Dutch student. Once they become operational, ambulance drones could reach heart attack patients in minutes and potentially save thousands of lives [5].

出 处

1. Tiny airplanes and subs from University of Florida laboratory could be next hurricane hunters, http://goo.gl/6aHTz6
2. DRONES MAPPED THIS GIANT VOLCANO, AND THE 3D RENDERING IS AWESOME, http://goo.gl/hXRKe1
3. CONSERVATION DRONE PROJECT, http://goo.gl/0K5OSK
4. Despite FAA dithering, a drone economy sprouts on the farm, http://goo.gl/7QTFDn
5. This Ambulance Drone Can Fly Into Trouble With First Aid, http://goo.gl/qTz18y

核心语法

熟悉 # 句式 5〈主语 +help+ 宾语 + 宾补〉。例文将介绍 # 宾语和 # 宾补是〈主语 + 谓语〉的句子。help 是动词，后接 # 动词原形或者 # 不定式 to 作 # 宾补。

- Their reports could **help** scientists [**to**] **improve** their storm–prediction powers. 它们的报告可以帮助/科学家提高/预测台风的能力。
- The model will **help** scientists [**to**] **understand** volcanoes. 模型可以帮助/科学家理解/火山。
- Drones with digital cameras **help** farmers [**to**] **identify** problem areas. 带数码相机的无人机可以帮助/农民发现/问题区域。

熟悉 # 不定式 to 作 # 副词的用法。此处意为“为了做 ~”。

- The team is using drones **to track** tropical storms. 这个团队在使用无人机/以追踪热带风暴。
- Geologists used special software **to render** a 3D model. 地质学家使用特殊软件/以建立3D模型。
- Farmers are using drones **to monitor** fields. 农民正使用无人机/监控田地。

熟悉 # 并列句。# 并列句冗长且复杂，但理解原理后并不难。将表对等的内容放在一起，最后加 # 并列连词 and、or、but 等。

- Drones can **fly** into the heart of a storm **and spy** on it. 无人机可以飞入风暴中心/并监测它。
- The drones **are launched** by a laptop, **use** minimal power, **and can be carried** by wind. 无人机由笔记本电脑发射/它们耗电量小/并且可被风带动运转。
- They use the military drones to **monitor** wildlife populations **or map** roads and wetlands. 他们使用军事无人机/监测野生动物数量或者测绘道路和湿地。
- Farmers are using drones to **monitor** fields, **increase** yields, **and save** money. 农民使用无人机/监测农田/提高产量/节省费用。

单词&短语

delivery 投递，寄送
lead 引导
discussion 讨论
unmanned 无人的
aerial 航空的
vehicle 坐骑
long before ~ 早在 ~ 之前
a swarm of ~ 一群 ~
track 追踪
tropical storm 热带风暴 launch 发射
minimal power 最低电量
water current 水流
massive 巨大的
temperature 温度
Pressure 压力
humidity 湿度
prediction 预测
thousands of 数千的
piece 连接 fascinating 迷惑性的
geologist 地质学家
be full of 充满
render 制作
wildlife 野生动物
raven 乌鸦
population 人口
wetland 湿地
conservancy 管理
identify 确认
distribution 分布
density 密度
field 田地，领域
yield 产量
close inspection 密切检测
infrared light 红外线
photosynthetic efficiency 光合效率
first responder 急救人员
search-and-rescue 搜救
lifeguard 救生员
prototype 样机
Dutch 荷兰的
operational 运转之前
heart attack patient 心脏麻痹患者
in minutes 在几分钟内
potentially 潜在地

技术术语

UAV 无人航空器以“无人机”的名称广为人知，无人搭乘，可远程操控。广义上可分为无人驾驶飞行器和远程遥控飞行器。因为无人搭乘，所以 UAV 经常用于从事辛苦且危险的工作。

根据提示完成句子

When people talk about drones, /

当人们谈论无人机时，/

Amazon's ________ ________ ______ _____ ____ ________. /

亚马逊的配送无人机总是首当其冲被谈及。/

However, / drones, or ________ ______ _______ (UAVs), / were used /

但是，/ 无人机或者无人航空器/被使用/

____ ______ Jeff Bezos __________ ___ ____. /

早在杰夫·贝佐斯发布他的计划之前。/

Originally drones were used /

最初无人机使用/

for _______ and ____________ applications / like _____ and ________. /

是为了军事和谍报应用/ 诸如监听和轰炸。/

Today ____ ___ ____ _________ drones / are being used /

现在越来越多的民间无人机被使用/

in _________ _____. / _____ ___________ are also ________ ____ /

在一些惊人的领域。/ 救生员也获得了帮助/

from drones. / _____-and-______ drones can fly /

从无人机。/ 搜救无人机可以飞越/

over __________ _____ / and search for __________. /

危险的地区/ 搜寻幸存者。/

Lifeguard drones reach ___________ / faster than _____ __________. /

救生无人机可到达泳客身边/ 比救生员更迅速。/

____ ____ / a _________ of an ambulance drone was ___________ /

去年/ 急救无人机的样机被介绍/

by a Dutch student. / Once they ______ ____________, /

由荷兰的一个学生。/ 一旦他们开始运行，/

ambulance drones could reach _____ _______ _________ /

急救无人机可以迅速到达心脏麻痹患者身边/

__ ________ / and _____________ ____ _________ __ _____.

在几分钟内/ 并且能潜在挽救数千人的生命。

思考题

解答题

1 （理解）以下哪一项不是无人机的应用领域？

ⓐ 台风调查　　ⓑ 扑灭山火

ⓒ 保护野生动物　　ⓓ 农事

2 （理解）以下哪一项没有正确阐述无人机？

ⓐ 可接近台风

ⓑ 调查野生动物的移动路径和分布

ⓒ 急救无人机可拯救心脏麻痹患者

ⓓ 使用紫外线照相机调查光合作用效率。

3 （论述）请简述无人机设计初衷。

4 （论述）请列举可救人的无人机。

讨论

1 亚马逊计划用无人机做什么？（提示：亚马逊是全球最大电商之一）

2 使用无人机拍摄高清数码照片时会出现什么问题？

课堂总结

1 学习#句式5、#使役动词、#感官动词的概念和例句。

2 学习#不定式to作#副词的例句。

3 学习#并列句、#并列连词、#关联连词概念和例句。

答案

1 b.（目前为止还不适用于扑灭山火） 2 d.（使用红外线相机） 3 军事和间谍目的（监听和轰炸） 4 分布在山间的搜救无人机、海边的救生无人机、城市中心的急救无人机

翻译

Five Cool Uses for Drones

无人机的五种特色用途

When people talk about drones, / Amazon's delivery drones often lead the discussion. /
当人们谈论无人机时，/ 亚马逊的配送无人机总是首当其冲被谈及。/
However, / drones, or unmanned aerial vehicles (UAVs), / were used /
但是，/ 无人机或者无人航空器 / 被使用 /
long before Jeff Bezos announced his plan. / Originally drones were used /
早在杰夫·贝佐斯发布他的计划之前。/ 最初无人机被用于 /
for military and intelligence applications / like spying and bombing. /
军事和谍报应用 / 诸如监听和轰炸。/
Today more and more civilian drones / are being used / in surprising areas. /
现在越来越多的民间无人机 / 被用 / 在一些惊人的领域。/

Here are five cool uses for drones: /
以下为无人机的 5 种用途。/

Hurricane Hunting / 台风袭击

Drones can fly / into the heart of a storm / and spy on it / as the storm develops. /
无人机可以飞入 / 风暴中心 / 并且监测它 / 在风暴进化过程中。/
A team at the University of Florida / is using a swarm of six-inch-long drones /
弗洛里达大学的一个团队 / 使用 6 英寸长的无人机群，/
to track tropical storms [1]. / The drones are launched / by a laptop, / use minimal power, /
以追踪热带风暴。/ 无人机被发射 / 通过笔记本电脑，/ 它们耗电量小，/
and can be carried / by wind or water current. / They ride / through a massive hurricane /
可被 / 风或者水流带动运转。/ 它们穿越 / 巨大的台风 /
and collect data / such as temperature, pressure, humidity, and location. /
并且收集数据 / 例如温度、压力、湿度、位置。/
Their reports could help scientists / improve their storm-prediction powers. /
它们的报告可以帮助科学家 / 提高预测台风的能力。/

3D Mapping / **3D**制图

Small, lightweight drones fly / over landscapes / and take thousands of digital images. /
轻量级无人机飞越 / 广阔的地域 / 并拍摄数千张数码照片。/
Then software pieces them / into 3D maps. / A 3D model of the Ambrym Volcano /
然后软件将这些碎片 / 拼成 3D 地图。/ 安布里姆岛马鲁姆火山的 3D 模型 /
on the Pacific island of Vanuatu / is a fascinating example [2]. /
位于太平洋瓦努阿图岛 / 是一个梦幻般的例子。/

The volcano was deeper / than the Empire State Building / and full of hot lava! /
火山 / 比帝国大厦要高 / 布满了滚烫的熔岩！ /
Geologists used a drone / and special drone mapping software / to render the 3D model. /
地质学家使用无人机 / 和特殊的无人机地图软件 / 构建 3D 模型。/
The model will help scientists / understand volcanoes and their risks. /
模型帮助科学家 / 了解火山及其危险性。/

Protecting Wildlife / 野生动物保护

The U.S. government already uses drones / to protect its lands and wildlife. /
美国政府已经使用无人机 / 保护领土和野生动物。/
Several agencies are using the military Raven drones / to monitor wildlife populations /
很多机构正在使用军事无人机 / 监测野生动物数量 /
or map roads and wetlands / for land management purposes. /
或者测绘道路和湿地 / 目的是进行土地管理。/
An Orangutan Conservancy also uses drones / to identify the animals' distribution and density /
一家猩猩管理机构也利用无人机 / 确认动物的分布和密度 /
in Indonesia and Malaysia [3]. /
在印度尼西亚和马来西亚。/

Farming / 农事

Farmers are using drones / to monitor fields, / increase yields, / and save money. /
农民使用无人机 / 监测农田，/ 提高产量，/ 节省成本。/
Drones with digital cameras help farmers / identify problem areas / for closer inspection, /
携带数码相机的无人机帮助农民 / 确认问题区域，/ 为实现密切检测。/
but now drones / with infrared light cameras / can inspect crop health /
但现在无人机 / 带有红外线相机的 / 可以调查农作物的健康状况 /
by detecting photosynthetic efficiency / in various plants [4]. /
通过侦测光合作用效率 / 在各种植物中的。/

Search and Rescue / 搜救

First responders are also getting help / from drones. / Search-and-rescue drones can fly /
急救人员一直获得帮助 / 从无人机。/ 搜救无人机可以飞越 /
over dangerous areas / and search for survivors. / Lifeguard drones reach swimmers /
危险区域 / 搜救幸存者。/ 救生无人机可到达泳客身边 /
faster than human lifeguards. / Last year / a prototype of an ambulance drone was introduced /
比救生员更快。/ 去年 / 急救无人机的样机被介绍 /
by a Dutch student. / Once they become operational, /
由荷兰的一个学生。/ 一旦他们开始运行，/
ambulance drones could reach heart attack patients /
急救无人机可以迅速到达心脏麻痹患者身边 /
in minutes / and potentially save thousands of lives [5]. /
在几分钟内 / 并能潜在挽救数千人的生命。/

机器人记者的崛起
The Rise of Robot Journalists

Robots are becoming smarter. Now they are writing news articles for you. You may think you'd be able to tell if an article is written by a robot. You are wrong! Most of us wouldn't notice at all. See for yourself!

Los Angeles Times' robot journalist **Quakebot** writes about earthquakes in Southern California. The robot automatically writes blog posts and even tweets. This is the first article Quakerbot wrote, on March 17, 2014 [1].

> A shallow magnitude 4.7 earthquake was reported Monday morning five miles from Westwood, California, according to the U.S. Geological Survey. The temblor occurred at 6:25 am. Pacific time at a depth of 5.0 miles.
>
> According to the USGS, the epicenter was six miles from Beverly Hills, California, seven miles from Universal City, California, seven miles from Santa Monica, California and 348 miles from Sacramento, California. In the past ten days, there have been no earthquakes of magnitude 3.0 and greater centered nearby.

Associated Press published a story titled "Apple tops Street 1Q forecasts", immediately after Apple released its first quarter earnings [2]. Yes, the story is written by a robot called **WordSmith**.

> The results surpassed Wall Street expectations. The average estimate of analysts surveyed by Zacks Investment Research was for earnings of $2.60 per share.
> The maker of iPhones, iPads and other products posted revenue of $74.6 billion in the period, also exceeding Street forecasts. Analysts expected $67.38 billion, according to Zacks.

WordSmith now produces 4,400 earnings stories per quarter, more than ten times the number AP's human reporters produce.

Forbes.com carries stories with the byline of **Narrative Science** [3]. Narrative Science is, according to Forbes.com, an artificial intelligence platform that transforms data into stories and insights. It's just a fancy description for a robot. Here's what Narrative Science wrote:

> Analysts are looking for decreased profit for Koppers Holdings when the company reports its results for the fourth quarter on Thursday, February 26, 2015. Koppers Holdings reported profit of 44 cents a year ago, but the consensus estimate calls for earnings per share of 40 cents this time around.

Now do you think you can tell robot-written stories from human-written ones?

出处

1. The First News Report on the L.A. Earthquake Was Written by a Robot, http://goo.gl/5NRZeT
2. Apple Tops Street 1Q Forecasts, http://goo.gl/BqmRJa
3. Thought of The Day, http://goo.gl/Hdcz1l

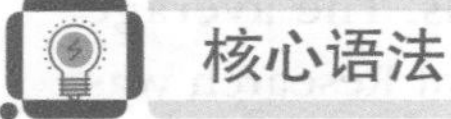

核心语法

熟悉 # 倍数的表达。意为"~的~倍"，不能与 of 同时使用。

- Her salary is **ten times ~~of~~ my salary**. 她的工资是我的10倍。
- It is more than **ten times ~~of~~ the number** of human-written stories. 这是人类撰写报道数量的10倍以上。

熟悉 # 关系代名词的省略。# 宾格关系代名词、# 主格关系代名词 +be 动词省略后不影响句意。

- This is **the first article** ([**that**] the robot wrote). 这是首篇报道/机器人写的。
- There have been no **earthquakes** of magnitude 3.0 and greater ([**which** were] centered nearby). 没有3.0级及以上的地震/在震源附近。
- AP published **a story** ([**which** was] titled "Apple tops Street 1Q forecasts.") AP发布了一篇报道/题为Apple tops Street 1Q forecast。
- The average estimate of **analysts** ([**who** were] surveyed) was for earnings of $2.60 per share. 分析人士预测/被调查的/每周平均纯利润为2.6美元。
- WordSmith produces **4,400 stories** per quarter, [**which** is] more than ten times **the number** ([**that**] human reporters produce). WordSmith写出4 400 篇报道，/每季度/这是10倍以上/数量的/人类记者创作的。

新闻阅读提示。

- According to the USGS, the epicenter was six miles from Beverly Hills, California, seven miles from Universal City, California, seven miles from Santa Monica, California and 348 miles from Sacramento, California.

  ☞ According to the USGS, the epicenter was six miles from **A**, seven miles from **B**, seven miles from **C** and 348 miles from **D**. 据美国地质调查局称，震源距离A地6英里、距离B地7英里、距离C地7英里、距离D地348英里。

▶ 地名按照行政区划由小到大的顺序排列，Bevery Hills和California分别指比弗利山庄和加利福尼亚州。句子夹杂很多大写字母和逗号或者结构较复杂时，可用简单的符号代替地名。

● Analysts are looking for decreased profit for Koppers Holdings when the company reports its results for the fourth quarter on Thursday, February 26, 2015.

☞ Analysts are looking for decreased profit for **A** when the company reports its results for the fourth quarter on **WHEN**. 分析人士期待/A的纯利润下降/公司汇报/结果时/第四季度的/当。

▶ 句子夹杂很多固有名词、日期、大写字母和逗号时，一般很难迅速理解含义。此时可以化难为易，用简单的符号代替上述内容。

单词&短语

news article 新闻报道
tell 区分
notice 告知
see for yourself 亲自确认
journalist 记者
earthquake 地震
automatically 自动地
shallow 浅的，弱的
magnitude 大小
geological survey 地质调查
temblor 地震
occur 发生
at a depth of ~ 在深度 ~
according to ~ 按照 ~
epicenter 震源
nearby 附近
titled 题为 ~
immediately after ~ 在 ~ 之后
release 公开，发布
the first quarter earning 第一季度收益
called ~ 称为 ~
surpass 超越
expectation 期待，期待值
average estimate 平均预估
analyst 分析人士
survey 调查
investment 投资
per share 每股
revenue 收益
exceed 超过
forecast 预测
produce 生产
byline（报纸、杂志的）署名
artificial intelligence platform 人工智能平台
transform A into B 将 A 转换为 B
insight 见解
fancy 帅气的、很棒的
description 说明，描述
look for 寻找，期待
decreased profit 收益减少
consensus 协商
call for 邀请，请求
tell A from B 从 B 中将 A 区分

根据提示完成句子

A ________ __________ 4.7 __________ was ________ /
震级为4.7级的浅源地震被报道/

Monday morning / five miles from Westwood, California, /
在周一早晨/ 距离加利福尼亚州西木区5英里，/

__________ __ the U.S. Geological Survey. / The temblor occurred /
据美国地质调查局。/ 地震发生于/

at 6:25 a.m. Pacific time / at a depth of 5.0 miles.
太平洋时间上午6：25/ 震源深度5英里。/

__________ __ the USGS, /
据美国地质调查局称，/

the _________ was six miles from Beverly Hills, California, /
震源距离加利福尼亚州比弗利山庄6英里，/

seven miles from Universal City, California, /
距加利福尼亚环球影城7英里，/

seven miles from Santa Monica, California /
距加利福尼亚圣莫妮卡7英里，/

and 348 miles from Sacramento, California. /
距加利福尼亚萨克拉门托348英里。/

In the past ten days, /
在过去10天里，/

there have been no ____________ of __________ 3.0 and _______ /
震级为3.0级及以上的地震没有发生过/

________ ______. /
在震源附近。/

Analysts are ________ ___ __________ ______ / for Koppers Holdings /
分析人士期待纯利润的下降/ Koppers Holdings的/

when the company reports its results / for the ______ _______ /
当公司公布结果/ 第四季度的/

on Thursday, February 26, 2015. /
在2015年2月26日周四。/

Koppers Holdings reported ______ of 44 cents / a year ago, /
Koppers Holdings公布其利润为44美分/ 1年前，/

but the ________ _________ _____ ___ __________ ___ _____ of 40 cents /
但调查结果预计，/每股利润会降至40美分/

this time around. /
此次。

思考题

讨论

1 请回答《纽约时报》在Did a Human or a Computer Write This?（http://goo.gl/MGTY09）网页上提出的问题。

2 未来，人类记者这个职业会完全消失吗？如果不消失，人类记者将主要负责何种领域？

3 机器人记者的算法是如何实现的？（提示：https://goo.gl/ElwDxA）

课堂总结

1 学习#比较句、#倍数、#分数的表达和例句。

2 学习#限定性定语从句和#非限定性定语从句。

3 学习#关系代名词的省略和例句。

翻译

The Rise of Robot Journalists

机器人记者的崛起

Robots are becoming smarter. / Now / they are writing news articles / for you. / You may think /
机器人愈来愈智能。/ 现在 / 它们可以撰写报道 / 为你。/ 你可能认为 /
you'd be able to tell / if an article is written / by a robot. / You are wrong! / Most of us wouldn't notice /
自己可以区分 / 报道是否 / 出自机器人之手。/ 你错了！ / 我们多数人都无法分辨 /
at all. / See for yourself! /
完全。/ 你自己看看！ /

Los Angeles Times' robot journalist Quakebot writes about earthquakes / in Southern California. /
《洛杉矶时报》的机器人记者Quakebot撰写了关于地震的报道/发生在南加利福尼亚的。/
The robot automatically writes blog posts / and even tweets. / This is the first article / Quakerbot wrote, /
机器人自动撰写博客报道 / 甚至推文。/ 这是首篇报道 /Quakerbot 撰写的， /
on March 17, 2014 [1]. /
在 2014 年 3 月 17 日。/

> A shallow magnitude 4.7 earthquake was reported / Monday morning /
> 震级为4.7级的浅源地震被报道/在周一早晨/
> five miles from Westwood, California, / according to the U.S. Geological Survey. /
> 距离加利福尼亚西木区5英里的，/据美国地质调查局。/
> The temblor occurred / at 6:25 a.m. Pacific time / at a depth of 5.0 miles.
> 地震发生于/太平洋时间上午6：25/震源深度5英里。/
>
> According to the USGS, / the epicenter was six miles from Beverly Hills, California, /
> 据美国地质调查局称，/震源距离加利福尼亚比弗利山庄6英里，/
> seven miles from Universal City, California, / seven miles from Santa Monica, California /
> 距加利福尼亚环球影城7英里，/距加利福尼亚圣莫妮卡7英里，/
> and 348 miles from Sacramento, California. / In the past ten days, /
> 距加利福尼亚萨克拉门托348英里。/在过去10天里，/
> there have been no earthquakes of magnitude 3.0 and greater / centered nearby. /
> 震级为3.0级及以上的地震没有发生过/在震源附近。/

Associated Press published a story / titled "Apple tops Street 1Q forecasts", /
美联社发布了一篇报道 / 题为 Apple tops Street 1Q forecasts， /
immediately after Apple released its first quarter earnings [2]. /
在苹果公司公布第一季度利润之后。/

Yes, the story is written / by a robot called WordSmith. /
是的，/ 这篇报道被撰写 / 由一个名为 WordSmith 的机器人。/

> The results surpassed Wall Street expectations. /
> 结果超出了华尔街的期待。/
> The average estimate of analysts / surveyed /
> 分析人士普遍估计/被调查的/
> by Zacks Investment Research / was for earnings of $2.60 per share. /
> 由扎克斯投资研究所进行的/每周平均纯利润2.6美元。/
> The maker of iPhones, iPads and other products posted revenue of $74.6 billion /
> iPhone、iPad和其他产品的制造商公布其销售额为746亿美元/
> in the period, / also exceeding Street forecasts. / Analysts expected $67.38 billion, /
> 在此期间，/也超出了华尔街的预测。/分析人士的期待值为673亿8000万美元，/
> according to Zacks. /
> 据扎克斯投资研究所。/

WordSmith now produces 4,400 earnings stories / per quarter, / more than ten times /
WordSmith 现在撰写 4400 篇报道，/ 每季度，/ 是 10 倍以上 /
the number AP's human reporters produce. /
美联社人类记者撰写数量的。/

Forbes.com carries stories / with the byline of Narrative Science [3]. /
福布斯刊登的报道 / 带有 Narrative Science 的署名。/
Narrative Science is, / according to Forbes.com, / an artificial intelligence platform /
Narrative Science 是，/ 据福布斯称，/ 人工智能平台 /
that transforms data into stories and insights. / It's just a fancy description / for a robot. /
它将数据转换为报道和见解。/ 这只是一个很棒的说明 / 关于机器人的。/
Here's what Narrative Science wrote: /
以下为 Narrative Science 撰写的内容：/

> Analysts are looking for decreased profit / for Koppers Holdings /
> 分析人士期待纯利润的下降/关于Koppers Holdings的/
> when the company reports its results /
> 当公司公布结果/
> for the fourth quarter / on Thursday, February 26, 2015. /
> 第四季度的/在2015年2月26日周四。/
> Koppers Holdings reported profit of 44 cents / a year ago,/
> Koppers Holdings公布其利润为44美分/1年前，/
> but the consensus estimate calls for earnings per share of 40 cents / this time around. /
> 但调查结果预计每股利润会降至40美分/此次。/

Now do you think / you can tell robot-written stories / from human-written ones? /
现在你认为 / 自己能区分机器人撰写的报道 / 和人类撰写的报道吗？/

机器人比人类工作更出色！

No Humans Please, Robots Do Better!

Will robots take our jobs? That was the BIG question in 2014. Opinions from 2,000 experts are divided, according to a report by Pew Research [1]. 52% claim robots will create more jobs while 48% worry robots can lead to mass unemployment. Another study by Oxford University predicts nearly half of US jobs are at risk over the next couple of decades [2]. That sounds scary!

Well, if you are concerned about jobs in the future, think again. Robots are already working among us. In fact, they are better workers than you at many tasks! Here are just five examples among many.

Warehouse worker

In several Amazon warehouses, or fulfilment centers, 15,000 Kiva robots are working on shipping orders [3]. Now, warehouse workers don't walk around, find the right shelf, and pick items out from the shelf. Instead, robots bring the right shelves to the workers. The workers just pick items out from the delivered shelves. The robot-equipped warehouses process at least 300 items per hour, three times more than the old system.

Soldier

Robots could replace one-fourth of all U.S. combat soldiers by 2030, according to U.S. Army General Robert Cone [4]. But the U.S. Military is already using robot soldiers for risky and dangerous jobs. Robots are scouring caves in Afghanistan, digging up roadside bombs in Iraq, and guarding the borders in war zones.

Pharmacist

Medication dispensing errors, or giving wrong pills, account for approximately 21% of all medical errors. At a pharmacy at the University of California, a robot pharmacist filled as many as 350,000 doses of medicine without a single error. Currently, at more than a third of major hospital pharmacies in the United States, robots fill over 350 million doses each year with a 99.9% accuracy rate.

Farmer

Farming involves a lot of routine tasks, and robots are perfect for them. In fact, "farmbots" are already working in the field [5]. A "lettuce bot" pulls up weeds. A "wine bot" prunes vines in vineyards. Other sensor-equipped bots monitor crops and test soil. Many dairy farms are using milking robots. Farm robots can work all day and night without breaks.

Housekeeper

Many households around the globe are already using housekeeping robots. The world famous vacuuming robot, the Roomba, is just one example. According to Roomba's maker, iRobot, the company has sold more than 10 million home robots worldwide.

出 处

1. The Web at 25, http://goo.gl/TgZwyD
2. The Future of Employment: How susceptible are jobs to computerisation? http://goo.gl/kQCyd8
3. Meet the Robots Shipping Your Amazon Orders, http://goo.gl/xdmDU2
4. U.S. Army general says robots could replace one-fourth of combat soldiers by 2030, http://goo.gl/PTOUai
5. 5 Coolest Farm Robots, http://goo.gl/LsBWFI

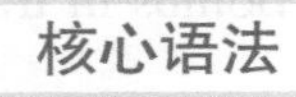

核心语法

熟悉 # 现在进行时态〈be+ 现在分词〉。意为“正在做 ~”。

- Robots **are** already **working** among us. 机器人正在我们周围工作。
- Robots **are scouring** caves in Afghanistan. 机器人在侦测洞穴/在阿富汗。
- Many dairy farms **are using** milking robots. 很多奶牛场在使用挤奶机器人。

熟悉 # 句式 3〈主语 + 动词 + 宾语〉中 #that 引导的从句作 # 宾语的句子。#that 引导的从句作 # 宾语时，that 可以省略。

- 52% claim [**that**] robots will create more jobs. 52%的人认为/机器人会创造更多工作机会。
- 48% worry [**that**] robots can lead to mass unemployment. 48%的人担心/机器人会让很多人失业。
- Another study predicts [**that**] nearly half of US jobs are at risk. 另一项研究预测/将有一半的美国职业濒临消失的危险。

熟悉 # 比较句。

- Robots are **better** workers **than** you at many tasks. 机器人比你更出色/在很多工作上。
- The robot-equipped warehouses process **three times more than** the old system. 配有机器人的物流仓库处理速度/比旧系统快3倍。
- A robot pharmacist filled **as many as** 350,000 doses of medicine. 机器人药剂师调配/高达350 000剂量的药。

熟悉 # 并列句。# 并列句看似冗长且复杂，但其实很简单。连接表并列的单词、从句、句子时，只需在最后加入 # 并列连词 and、or、but 等即可。

- Warehouse workers don't **walk** around, **find** the right shelf, **and pick** items out from the shelf. 仓库工人不再到处走动，不再寻找正确的货架，也不再从中挑选货物。
- Robots are **scouring** caves in Afghanistan, **digging** up roadside bombs in Iraq, **and guarding** the borders in war zones. 机器人在阿富汗侦测洞穴，在伊拉克清除炸弹，在战区守卫边界线。
- Other robots **monitor** crops **and test** soil. 其他机器人监测作物和测试土壤。

单词&短语

opinion 意见
lead to 导致 ~
mass unemployment 大量失业
nearly half of ~ 接近一半的 ~
be at risk 濒临危险
be concerned about 担心 ~
fulfillment center 物流中心
warehouse worker 仓库职员
walk around 四处走动
pick out 挑选
deliver 寄送，运送
-equipped 配备 ~
replace 替换
scour 侦查
cave 洞穴
dig up 挖掘，开采
roadside bomb 路边炸弹
border 国境，边界线
medication dispensing error 配药出错
account for 说明，成为原因
approximately 大量的，约
pharmacy 药店
pharmacist 药剂师
dose（药物的）剂量
accuracy rate 准确度
involve 包括，涉及
perfect for 非常适合 ~
farmbot 农活好帮手
pull up weeds 除草
prune vines 修剪葡萄藤
vineyard 葡萄园
crop 作物
soil 土壤
dairy farm 奶牛场
milking robot 挤奶机器人
day and night 不分昼夜
without a break 不休息
household 家政
routine task 日常工作
vacuum 真空，用真空吸尘器打扫
worldwide 世界范围内

技术术语

Kiva 搬运货物以实现仓库自动化的机器人。它依托无线网络和基于服务器的后端系统，根据无线指令的订单将货物所在的货架从仓库搬运至员工处理区，由员工最终打包。

根据提示完成句子

Will ______ ____ ___ ____? / That was the BIG question / in 2014. /
机器人会抢走我们的工作吗？/ 这是一个重大问题/ 在2014年。/

______ ____ _____ _______ ___ ________ / __________ __ _ _______ /
2000位专家的意见被划分，/ 根据报告/

by Pew Research [1]. / 52% _____ / ______ ____ ______ _______ /
皮尤研究中心的。/ 52%的专家认为/ 机器人会创造更多工作/

while 48% ______ / ______ ___ ____ __ ____ ___________. /
而48%的专家担心/ 机器人会导致大量失业。/

Another _____ by Oxford University /
另一项牛津大学的调查/

______ ______ ____ __ __ ____ ___ __ ____ /
预计/将近一半的美国职业濒临消失的危险/

over the ____ ______ __ _______ [2]. / That ______ _____!
在未来的20~30年内。/这听起来很惊悚！/

In several Amazon ___________, / or __________ centers, /
在众多亚马逊仓库中，/或者物流中心中，/

15,000 Kiva robots are working / on ________ ______ [3]. /
15 000台Kiva机器人在工作/在装载命令下。/

Now, __________ _______ don't ____ _____, / ____ ___ _____ _____, /
现在，仓库职员不再到处走动，/ 寻找正确的货架，

and ____ _____ / out from the shelf. / Instead, /
挑选货物/ 从货架中。/ 相反，/

robots bring the _____ _______ / to the workers. /
机器人会将正确的货架送到面前/ 工人的。

The workers just ____ _____ / out from the _________ _______. /
工人仅挑选货物/ 从运来的货架中。/

The _____-_______ __________ _______ /
配有机器人的物流仓库处理/

at least 300 items per hour, / three times more than the old system. /
每小时至少300件货物，/是旧系统的3倍。

思考题

解答题

1 （理解）机器人尚未介入以下哪一项工作？

ⓐ 军人　　ⓑ 农民

ⓒ 药剂师　　ⓓ 音乐家

2 （理解）以下哪一项没有正确描述机器人？

ⓐ 机器人的存在让员工变得多余。

ⓑ 使用机器人之后，配药准确率高达 99.9%。

ⓒ 机器人在军队从事危险艰难的工作。

ⓓ 家用机器人全球售出 1000 万台以上。

3 （论述）列举例文中机器人擅长的日常工作（重复性工作）。

4 （论述）预测未来20~30年内，机器人的存在会导致美国多少职业消失？

讨论

1 如果军用机器人在战斗中失误，该如何问责？

2 最近一些专业性的工作（律师、会计师、医生、法务）也受到了机器人的威胁，请列举只有人类可从事的工作。

课堂总结

1 学习#现在分词、#一般时态、#进行时态的概念和例句。

2 学习#动词后接#that从句作#宾语的例句。

3 学习使用#原级、#比较级、#最高级的#比较句。

答案

1 d. 2 a. 3 药物分类、农作物分类、清扫地板、货物分类 4 约 50%（一半）

翻译

No Humans Please, / Robots Do Better!

机器人比人类工作更出色!

Will robots take our jobs? / That was the BIG question / in 2014. /
机器人会抢走我们的工作吗? / 这是一个重大问题 / 在 2014 年。/
Opinions from 2,000 experts are divided, / according to a report / by Pew Research [1]. /
2000 位专家的意见被划分，/ 根据报告 / 皮尤研究中心的。/
52% claim / robots will create more jobs / while 48% worry / robots can lead to mass unemployment. /
52% 的专家认为 / 机器人会创造更多工作 / 而 48% 的专家担心 / 机器人会导致大量失业。/
Another study by Oxford University / predicts / nearly half of US jobs are at risk /
另一项牛津大学的调查 / 预计 / 将近一半的美国职业濒临消失的危险 /
over the next couple of decades [2]. / That sounds scary!
在未来 20~30 年内。/ 这听起来很惊悚！ /

Well, / if you are concerned about jobs / in the future, / think again. /
好吧，/ 如果你对工作深感忧虑 / 将来 / 再仔细考虑一下。/
Robots are already working / among us. / In fact, / they are better workers / than you / at many tasks! /
机器人已经开始工作 / 在我们周围。/ 事实上 / 它们是更好的工人 / 比你 / 在很多任务！ /
Here are just five examples / among many. /
以下仅是 5 个例子 / 在众多案例中。/

Warehouse worker / 仓库工人

In several Amazon warehouses, / or fulfilment centers, / 15,000 Kiva robots are working /
在众多亚马逊仓库中，/ 或者物流中心中，/15 000 台 Kiva 机器人在工作 /
on shipping orders [3]. / Now, warehouse workers don't walk around, / find the right shelf, /
在装载命令下。/ 现在，仓库职员不再到处走动，/ 寻找正确的货架，/
and pick items / out from the shelf. / Instead, / robots bring the right shelves / to the workers. /
挑选货物 / 从货架中。/ 相反，/ 机器人会将正确的货架送到面前 / 工人的。/
The workers just pick items / out from the delivered shelves. /
工人仅挑选货物 / 从运来的货架中。/
The robot-equipped warehouses process / at least 300 items per hour, /
配有机器人的物流仓库处理 / 每小时至少 300 件货物 /
three times more than the old system. /
是旧系统的 3 倍。

Soldier / 士兵

Robots could replace one-fourth of all U.S. combat soldiers / by 2030, /
机器人可取代全美国 1/4 的军人 / 截止到 2030，/

according to U.S. Army General Robert Cone [4]. /
据美国陆军上将罗伯特·科恩所言。/
But the U.S. Military is already using robot soldiers / for risky and dangerous jobs. /
但是美国陆军已经开始使用机器人士兵 / 从事危险和艰辛的工作。/
Robots are scouring caves in Afghanistan, / digging up roadside bombs / in Iraq, /
机器人在阿富汗侦测洞穴，/ 清除路边炸弹，/ 在伊拉克，/
and guarding the borders / in war zones. /
并且守卫边界 / 在战区。/

Pharmacist / 药剂师

Medication dispensing errors, / or giving wrong pills, /
配药错误，/ 或者取药错误，/
account for approximately 21% of all medical errors. /
是大约 21% 医疗事故的原因所在。/
At a pharmacy at the University of California, / a robot pharmacist filled /
在加利福尼亚大学的药店，/ 机器人配药 /
as many as 350,000 doses of medicine / without a single error. /
高达 350 000 剂 / 没有任何错误。/
Currently, / at more than a third of major hospital pharmacies / in the United States, /
现在，/1/3 以上的主要医院药店 / 美国的，/
robots fill over 350 million doses / each year / with a 99.9% accuracy rate. /
机器人配药高达 3 亿 5000 万次以上 / 每年 / 准确率为 99.9%。/

Farmer / 农民

Farming involves a lot of routine tasks, / and robots are perfect for them. /
农事涉及很多常规性工作 / 并且机器人很适合这些工作。/
In fact, / "farmbots" are already working / in the field [5]. / A "lettuce bot" pulls up weeds. /
事实上，/farmbots 已经开始工作 / 在农田。/lettuce bot 除草。/
A "wine bot" prunes vines in vineyards. / Other sensor-equipped bots monitor crops /
wine bot 机器人在葡萄园挖洞。/ 其他配备传感器的机器人监测作物 /
and test soil. / Many dairy farms are using milking robots. / Farm robots can work /
测试土壤。/ 很多奶牛场使用挤奶机器人。/ 农场机器人可以工作 /
all day and night / without breaks. /
昼夜 / 无休。/

Housekeeper / 家庭管家

Many households / around the globe / are already using housekeeping robots. /
很多家庭 / 全球范围内的 / 已经开始使用管家机器人。/
The world famous vacuuming robot, / the Roomba, / is just one example. /
世界最著名的真空清扫机器人，/Roomba，/ 仅是其中一例。/
According to Roomba's maker, / iRobot, / the company has sold /
据 Roomba 制造商说，/iRobot，/ 公司已经出售 /
more than 10 million home robots / worldwide. /
逾千万台家用机器人 / 在全球范围内。/

五大知名人士的忧虑

Five Brilliant Minds Are Concerned

Artificial intelligence (AI) is the intelligence exhibited by machines or software. The idea that a machine can be as intelligent as a human being has fascinated mankind for decades. Numerous science fiction novels and movies have explored the scenarios that may unfold once machines develop AI. Some are entertaining while others are frightening.

Now, in scientific circles, an increasing number of experts believe there is a reasonable chance that *the singularity* will happen. The singularity refers to the moment when machines become more intelligent than us. What will happen then? Five brilliant minds of our time are concerned. What about you? Are you worried?

Stephen Hawking

"The development of full artificial intelligence could spell the end of the human race. It would take off on its own and re-design itself at an ever increasing rate. Humans, who are limited by slow biological evolution, couldn't compete, and would be superseded," the world-renowned physicist said in an interview with the BBC [1].

Elon Musk

Musk is famous for his businesses on the cutting edge of technology, such as Tesla and SpaceX, yet he is concerned about AI. He warned that AI could be "the biggest existential threat" to mankind and said, "With artificial intelligence we are summoning the demon." He also tweeted that AI is “potentially more dangerous than nukes.”[2]

Bill Gates

Bill Gates wrote during an AMA (ask me anything) session on Reddit [3]: “I am in the camp that is concerned about super intelligence. First the machines will do a lot of jobs for us and not be super intelligent. That should be positive if we manage it well. A few decades after that, though, the intelligence is strong enough to be a concern. I agree with Elon Musk and some others on this and don’t understand why some people are not concerned.”

Vernor Vinge

Vinge, a mathematician and fiction writer who coined the term ‘the singularity’, believes the singularity is inevitable. “The competitive advantage — economic, military, even artistic — of every advance in automation is so compelling,” he wrote, “that passing laws, or having customs, that forbid such things merely assures that someone else will get them first.” What will happen when the singularity occurs? “The physical extinction of the human race is one possibility,” Vinge wrote.

Nick Bostrom

Bostrom, the philosopher and director of the Future of Humanity Institute at the University of Oxford, writes in his book *Superintelligence* that machines could eradicate humans with various strategies and that the world could become “a society of economic miracles and technological awesomeness, with nobody there to benefit. A Disneyland without children.”[4]

出 处

1. Stephen Hawking warns artificial intelligence could end mankind, http://goo.gl/1cQncG
2. Elon Musk’s Tweet, https://goo.gl/FwSUqv
3. BILL GATES FEARS A.I., BUT A.I. RESEARCHERS KNOW BETTER, http://goo.gl/sS6KCv
4. Superintelligence: Paths, Dangers, Strategies, by Nick Bostrom, http://goo.gl/1l0Hgb

核心语法

熟悉 # 情感类动词。此处着重强调 # 情感类动词是因为，其表现方式有别于中文，很容易将触发感情的主体和对象混淆。例如，在 He worries me. 句中，担心的人是“我”，担心的对象是“他”。例文将介绍采用相同表达方式的动词。

- Five brilliant minds **are concerned** about AI. 五大知名人士很担心AI。
- The idea **fascinated** mankind. 这个观点迷惑了人类。
  = Mankind **was fascinated** by the idea. =人类被这个观点迷惑了。
- Some scenarios are **entertaining** while others are **frightening**. 有些场景让人开心，而有些场景让人感到恐惧。

熟悉 #that 的诸多用法。不需要记住语法，理解句子原理即可。以下例句提及了语法术语，方便读者查找类似例句。

- The idea (**that** a machine can be as intelligent as a human being) has fascinated mankind.（机器和人类一样智能的）观点迷惑了人类。
  ▶ **#that为同位语**。that从句详细说明the idea。
- Numerous movies have explored the **scenarios** (**that** may unfold tomorrow). 很多电影已经探索出众多剧本/可能明天就会展现于世人面前。
  ▶ **#that为关系代名词**起到形容词的作用，修饰scenarios。
- He warned **that** AI could be the biggest threat to mankind. 他提醒/AI可能是人类最大的威胁。
  ▶ **#that引导宾语从句**。句子整体是warned的宾语。
- **That** should be positive if we manage it well. 这应该是积极的 / 如果我们认真管理。
  ▶ **#that为指示代名词**。代指前文出现的名词。
- Automation is so compelling **that** passing laws **that** forbid such things merely assures **that** someone else will get them first.
  = Automation is **so** compelling **that S** merely assures **O**. 自动化非常引人入胜/S只能保障O。

S = passing laws (**that** forbid such things) 通过法律/禁止此类事情的

O = **that** someone else will get them first 其他人会首先得到它们

▶ 第一个that是**#that引导副词从句**。so A that B可译为如此A以至于B。第二个that是**#that作关系代名词**。第三个that是assures的宾语，**#that引导宾语从句**。

单词&短语

exhibit 展示
as ~ as ... 和 ~ 一样……
fascinate 吸引
mankind 人类
decade 10 年
numerous 大量的
explore 探险
scenario 剧本
unfold 展开
once ~ 一旦 ~
entertaining 使人愉快的
frightening 令人恐怖的
scientific circles 科学界
a number of 众多的
increasing 增加
reasonable 合理的
Chance 机会，可能性
happen 发生
refer to 指的是
moment 时刻
brilliant 卓越的，优秀的
what about ~ ~ 怎么样?
development 发展，发达
spell 意味着
human race 人种
take off 起飞
at a(n) ~ rate 以 ~ 的速度
limited 有限的
biological evolution 生物学进化
compete 竞争
supersede 取代，代替
world-renowned 世界知名的
physicist 物理学家
be famous for ~ 因 ~ 出名
cutting edge 尖端
be concerned about ~ 担忧 ~
warn 警告
existential 存在主义的
threat 威胁
summon a demon 召唤恶魔
potentially 潜在地
nuke 核武器
in the camp ~ 和 ~ 是一伙的
positive 积极的
agree with 同意 ~
mathematician 数学家
coin 创造（单词）
inevitable 不可避免的
competitive advantage 竞争优势
economic 经济的
military 军事的
artistic 艺术的
advance 发展，进步
compelling 引人入胜的
pass a law 通过法令
customs 习惯
forbid 禁止
occur 发生
physical extinction 物理灭绝
possibility 可能性
philosopher 哲学家
eradicate 摧毁
strategy 战略
miracle 奇迹
awesomeness 可怕的
benefit 获取利益

技术术语

singularity 奇点指因文明演变而出现的未来假想点，急剧的技术变化极大地影响了未来，使人类生活无法回到过去。如果人类打造的超智能可以进行技术创造，那么迅猛的技术发展将超出人类掌控范围，今后将很难预测发生在人类身上的事情。详细内容请参考 http://goo.gl/Ntme2。

根据提示完成句子

Artificial intelligence (AI) is the intelligence / _________ /
AI是智能/ 被呈现的/

by machines or software. /
由机器或软件。/

The idea / that a machine can be as intelligent as a _____ _____ /
这个观点/说机器可以和人类一样智能的/

has fascinated mankind / ___ _______. /
迷惑了人类/ 数十年。/

Numerous science fiction novels and movies have explored the scenarios /
很多科幻小说和电影/已经探索出众多剧本/

that ___ ____ / once machines develop AI. /
未来可能发生的/ 一旦机器开发出AI。/

Some are ____________ / _____ others are ___________. /
有些场景让人开心，/ 而有些场景让人感到恐惧。/

Now, / __ _________ _______, / an increasing number of experts believe /
现在，/ 在科学界，/ 越来越多的专家相信/

there is a __________ ______ / that the ___________ will happen. /
有合理的可能性/ 奇点会出现。/

The ___________ ______ __ the moment /

这个瞬间/

when machines become more intelligent than us. /

机器比人类更智能。/

____ ____ ______ then? /

将会发生什么呢？/

Five _________ _____ of our ____ ___ _________. /

五位知名人士很担心AI。/

What about you? / Are you worried? /

你呢？/你也担心吗？/

思考题

讨论

1 你对人工智能期待更多还是忧虑更多？请简述立场。

2 请简述你对人工智能发展前景的期待。你认为类似《复仇者联盟 2》中“奥创”的人工智能可能实现吗？

课堂总结

1 学习#情感类动词的概念和例句。

2 学习#that的用法。

3 学习#关系代名词that的用法。

翻译

Five Brilliant Minds Are Concerned / 五大知名人士的忧虑

Artificial intelligence (AI) is the intelligence / exhibited / by machines or software. /
AI 是智能 / 被呈现的 / 通过机器或软件。/
The idea / that a machine can be as intelligent as a human being / has fascinated mankind / for decades. /
这个观点 / 说机器可以和人类一样智能的 / 迷惑了人类 / 数十年。/
Numerous science fiction novels and movies have explored the scenarios / that may unfold /
很多科幻小说和电影已经探索出众多剧本 / 可能发生的 /
once machines develop AI. / Some are entertaining / while others are frightening. /
一旦机器开发出 AI。/ 有些场景让人开心，/ 而有些场景让人感到恐惧。/

Now, / in scientific circles, / an increasing number of experts believe / there is a reasonable chance /
现在，/ 在科学界，/ 越来越多的专家相信 / 有合理的可能性 /
that the singularity will happen. / The singularity refers to the moment /
奇点将会出现。/ 奇点指这个瞬间 /
when machines become more intelligent than us. / What will happen then? /
机器比人类更智能的。/ 那时将会发生什么呢？ /
Five brilliant minds of our time are concerned. / What about you? / Are you worried? /
五大知名人士很担心 AI。/ 你呢？ / 你也担心吗？ /

Stephen Hawking / 斯蒂芬・霍金

“The development of full artificial intelligence could spell the end of the human race. /
“完全人工智能的开发意味着人类的灭绝。/
It would take off / on its own / and re-design itself / at an ever increasing rate. /
它会成功 / 自己 / 并且会重新设计自己 / 快速地。/
Humans, / who are limited / by slow biological evolution, / couldn’t compete, /
人类，/ 受限于 / 缓慢的生物进化 / 无法竞争，/
and would be superseded,” / the world-renowned physicist said / in an interview / with the BBC [1]. /
将被取代。” / 这位世界知名物理学家说 / 在采访 / 与 BBC 的。/

Elon Musk / 埃隆・马斯克

Musk is famous / for his businesses / on the cutting edge of technology, / such as Tesla and SpaceX, /
马斯克出名 / 因他的工作 / 关于尖端技术 / 例如特斯拉和 SpaceX，/
yet he is concerned about AI. / He warned / that AI could be “the biggest existential threat” to mankind /
但是他很担心 AI。/ 他警告称 /AI “可能成为人类最大的威胁” /
and said, “With artificial intelligence / we are summoning the demon.” / He also tweeted /
并说，“通过人工智能 / 我们正在召唤恶魔。” / 他也曾在 Twitter 发文，/

that AI is "potentially more dangerous / than nukes."[2] /
AI "内在更危险 / 比核武器"。/

Bill Gates / 比尔・盖茨

Bill Gates wrote / during an AMA (ask me anything) session / on Reddit [3]: / "I am in the camp /
比尔・盖茨写道 / 在 AMA（有问必答）会议期间 / 在 Reddit 上 / "我属于这个行列 /
that is concerned about super intelligence. / First the machines will do a lot of jobs / for us /
担心超智能的。/ 首先机器会做更多工作 / 为我们 /
and not be super intelligent. / That should be positive / if we manage it well. / A few decades after that, /
但超智能并不会。/ 这应该是积极的 / 如果我们妥善管理。/20~30 年后，/
though, / the intelligence is strong enough / to be a concern. / I agree with Elon Musk and some others /
但是，/ 如果智能足够强大 / 就值得担心。/ 我赞成埃隆・马斯克和其他人 /
on this / and don't understand / why some people are not concerned." /
在这件事情上 / 但是我不理解 / 为什么有些人不担心。" /

Vernor Vinge / 弗诺・文奇

Vinge, / a mathematician and fiction writer / who coined the term 'the singularity', /
文奇，/ 数学家、小说家 / 他创造了"奇点"这个术语 /
believes / the singularity is inevitable. /
相信 / 奇点是不可避免的。/
"The competitive advantage—economic, military, even artistic— / of every advance /
"竞争优势——经济、军事甚至艺术——/ 所有发展的 /
in automation / is so compelling," / he wrote, / "that passing laws, or having customs, /
在自动化中 / 如此有魅力，" / 他写道：/ "通过法令或者形成习惯，/
that forbid such things / merely assures / that someone else will get them first." /
阻止这种事情 / 只能保证 / 别人能率先得到他们。" /
What will happen / when the singularity occurs? /
将会发生什么 / 奇点出现时？ /
"The physical extinction of the human race is one possibility," / Vinge wrote. /
"人类物理灭绝是一种可能。" / 文奇写道。/

Nick Bostrom / 尼克・波斯特罗姆

Bostrom, / the philosopher and director of the Future of Humanity Institute / at the University of Oxford, /
波斯特罗姆 / 哲学家、人类未来研究所主任 / 在牛津大学，/
writes in his book Superintelligence / that machines could eradicate humans / with various strategies /
在他的著作《超级智能》中 / 机器会使人类灭亡 / 用各种策略 /
and that the world could become "a society of economic miracles and technological awesomeness, /
世界会成为"充满经济奇迹和技术恐慌的存在 /
with nobody there / to benefit. / A Disneyland without children."[4] /
无人 / 受益。/ 迪士尼乐园没有儿童。" /

经典语录之机器人篇

Notable Quotes on Robotics

Here are some notable quotes on Robotics and Artificial Intelligence. Some are from visionaries who shaped the future with their imagination.

Three Laws of Robotics are:

1. A robot may not injure a human being or, through inaction, allow a human being to come to harm.
2. A robot must obey the orders given it by human beings, except where such orders would conflict with the First Law.
3. A robot must protect its own existence as long as such protection does not conflict with the First or Second Law.

— **Issac Asimov** (1920~1992), author and professor at Boston University

"Man is a robot with defects."

— **Emile Cioran** (1911~1995), author of *The Trouble With Being Born*

"The question of whether a computer can think is no more interesting than the question of whether a submarine can swim."

— **Edsger W. Dijkstra** (1930~2002), computer scientist

"I believe that at the end of the century the use of words and general educated opinion will have altered so much that one will be able to speak of machines thinking without expecting to be contradicted."

— **Alan Turing** (1912~1954), pioneering computer scientist and mathematician

"Before we work on artificial intelligence, why don't we do something about natural stupidity?" — **Steve Polyak**, professor at University of Washington

"At bottom, robotics is about us. It is the discipline of emulating our lives, of wondering how we work."

— **Rod Grupen**, professor at University of Massachusetts Amherst

"We are survival machines—robot vehicles blindly programmed to preserve the selfish molecules known as genes."

— **Richard Dawkins**, author of *The Selfish Gene*

"I visualize a time when we will be to robots what dogs are to humans, and I'm rooting for the machines."

— **Claude Shannon**, author of *The Mathematical Theory of Communication*

"Some people worry that artificial intelligence will make us feel inferior, but then, anybody in his right mind should have an inferiority complex every time he looks at a flower." — **Alan Kay**, American scientist

"The sad thing about artificial intelligence is that it lacks artifice and therefore intelligence." — **Jean Baudrillard**, philosopher

"Intelligence is the art of good guesswork."

— **H. B. Barlow**, author of *The Oxford Companion to the Mind*

"Artificial Intelligence is a two-edged sword. On the one hand, it allows us to create intelligent artifacts with human-like perception and cognition. On the other hand, it accelerates people's heavy dependence on artifacts."

— **Max Bramer**, author of *Artificial Intelligence: An International Perspective*

"Today's AI is about new ways of connecting people to computers, people to knowledge, people to the physical world, and people to people."

— **Patrick Winston**, professor at MIT, MIT AI Lab briefing

核心语法

熟悉 # 动名词形式主语。# 动名词形式主语可以为 # 所有格和 # 宾格。# 宾格作主语的句子较难发现，建议反复参照例句学习。

- We will be able to speak of **machines thinking** without expecting to be contradicted. 我们可以谈论/机器思考/不用考虑会被反驳。

熟悉 # 关系代名词的省略。# 宾格关系代名词、# 主格关系代名词 +be 动词可以省略。省略后不影响原意。

- A robot must obey **the orders** [**which** are] given it by human beings. 机器人必须服从命令/由人类下达的。
- We are survival machines – robot vehicles [**that** are] blindly programmed to preserve the selfish **molecules** [**that** are] known as genes. 我们是幸存的机器——机器人工具/它被盲目地设计/以保护自私的分子/这些分子被称为基因。

熟悉 # 关系副词。# 关系副词连接两个句子，兼具 # 连词和 # 副词的作用。

- A robot must obey the orders, except **the case**. + Such orders would conflict with the First Law **in the case**. 机器人必须遵守命令，/除了一种情况 + 这种命令会违背第一定律/在这种情况下。

= A robot must obey the orders, except **the case in which** such orders would conflict with the First Law. =机器人必须遵守命令，/除了一种情况/这种命令会违背第一定律/在这种情况下。

= A robot must obey the orders, except **the case where** such orders would conflict with the First Law.

= A robot must obey the orders, except **where** such orders would conflict with the First Law.

▶ where是关系副词，指地点。地点可以是实际的地点，也可指情况、条件、状态等抽象的场所。原理与#关系代名词类似，但#关系副词和#先行词的省略条件不同，以下例句适用于两者。

● Anybody should have an inferiority complex **every time**. + He looks at a flower **at the time**. 每个人都会有自卑感/每次 + 他看花/的时候。

= Anybody should have an inferiority complex **every time when** he looks at a flower.

=每个人都会有自卑感/每当（每次）此时/他看花的。

= Anybody should have an inferiority complex **every time** he looks at a flower.

= Anybody should have an inferiority complex **whenever** he looks at a flower.

▶ every time when可用whenever替换。语法上称为#复合关系副词。

单词&短语

notable 值得注意的
artificial intelligence 人工智能
visionary 远见者
shape 使成型
imagination 想象，想象力
injure 伤害，受伤
inaction 无作为
obey an order 服从命令
conflict with 冲突，矛盾
protect 保护
existence 存在
as long as 只要 ~
defect 缺点
submarine 潜水艇
general 一般的
opinion 意见
alter 改变
speak of ~ 谈及 ~
contradict 反驳
natural 自然的，天生的
stupidity 愚蠢
at bottom 实际上的
discipline 学问
emulate 模仿
wonder 好奇
survival 幸存
vehicle 手段
blindly 盲目地
preserve 保护
selfish 自私的
molecule 分子
gene 基因
visualize 可视化，描绘，想像
A is to B what C is to D A 之于 B 相当于 C 之于 D
root for 支持
worry 担心
inferior 自卑的
inferiority complex 自卑感
artifice 技巧，策略
intelligence 智能
guesswork 推测

two-edged sword 双刃剑
on (the) one hand 一方面
on the other hand 另一方面
allow A to V 允许 A 做 V
perception 直觉，观察力
cognition 认知，认识
accelerate 加速
dependence on ~ 依赖于 ~
connect A to B 将 A 连接至 B
physical world 物质世界

根据提示完成句子

"Before we ____ __ __________ __________, /
"在我们致力于人工智能之前，/

why don't we do something about _______ _________?"
为什么不在天然愚蠢上做些事情呢？" /

— **Steve Polyak**, professor at University of Washington

"Artificial Intelligence is a ___-_____ sword. /
"人工智能是把双刃剑。/

__ ___ ___ ____, / it allows us /
一方面，/ 它允许我们/

to create intelligent artifacts with human-like _________ and _________. /
创造具备堪比人类感觉和意识的智能产品。/

__ ___ _____ ____, / it ___________ / people's _____ _________ /
另一方面，/ 它加速了/ 人类的过度依赖/

on _________." /
对人工产品的。"

— **Max Bramer**, author of *Artificial Intelligence: An International Perspective*

"Today's AI is about new ways / of _________ ______ to _________, /
"今日的AI是关于新方式/ 连接人类和计算机，/

______ to ________, / ______ to the ________ _____, /
人类和知识，/ 人类和物理世界，/

and ______ to ______." /
以及人与人的。" /

— **Patrick Winston**, professor at MIT, MIT AI Lab briefing

思考题

讨论

1 已通过"图灵测试"（可区分人工认知和机器认知）的人工智能预计何时问世？请简述主张及依据。

课堂总结

1 学习#动名词形式主语的概念和例句。
2 学习#关系代名词的概念和例句。
3 学习#关系副词和#复合关系副词的概念和例句。

翻译

Notable Quotes / on Robotics

经典语录之机器人篇

Here are some notable quotes / on Robotics and Artificial Intelligence. / Some are from visionaries /
以下是经典语录 / 关于机器人和人工智能的。/ 一些出自名人 /
who shaped the future / with their imagination. /
他们创造了未来 / 用自己的想象力。/

Three Laws of Robotics are / 机器人三定律

1. A robot may not injure a human being / or, through inaction, /
   机器人不得伤害人，/也不得袖手旁观/
   allow a human being to come to harm. /
   见人受到伤害。/
2. A robot must obey the orders / given it / by human beings, /
   机器人应服从一切命令/收到的/人下达的/
   except where such orders would conflict with the First Law. /
   但不得违反第一定律。/
3. A robot must protect its own existence /
   机器人应保护自身安全/
   as long as such protection does not conflict with the First or Second Law. /
   但不得违反第一、第二定律。/

——**艾萨克·阿西莫夫**（1920—1992），美国作家、波士顿大学教授

"Man is a robot with defects." ——**埃米尔·齐奥朗**（1911—1995），*The Trouble With Being Born* 作者
"人类是有缺陷的机器人。"/

"The question of whether a computer can think / is no more interesting /
"关于计算机会不会思考的问题 / 已经不再更引人关注 /
than the question of whether a submarine can swim." /
相比潜水艇会不会游泳。"/ ——**艾兹格·迪科斯彻**（1930—2002），计算机科学家

"I believe / that at the end of the century / the use of words and general educated opinion /
"我相信 / 在本世纪末 / 单词使用和一般受教育的观点 /
will have altered so much / that one will be able to speak of machines thinking /
将会有很大改变 / 以便我们将能够谈起机器思考 /
without expecting to be contradicted." /
而不用担心会遭到反驳。"

——**阿兰·图灵**（1912—1954），计算机科学先驱、数学家

"Before we work on artificial intelligence, / why don't we do something about natural stupidity?"
"在我们致力于人工智能之前，/ 为什么不在天然愚蠢上做些事情呢?"/

——**史蒂夫·波利亚克**，华盛顿大学教授

"At bottom, / robotics is about us. / It is the discipline of emulating our lives, /
"实际上，/ 机器人是关于我们的。/ 它模仿我们的生活，/
of wondering how we work." /
好奇我们的工作。"/

——**罗德·哥鲁本**，马萨诸塞大学安姆斯特分校教授

"We are survival machines – / robot vehicles / blindly programmed /
"我们是活着的机器——/ 机器人工具 / 被盲目地设计 /
to preserve the selfish molecules / known as genes."
以保存自私的分子 / 又称基因。"/

——**理查德·道金斯**，《自私的基因》作者

"I visualize / a time when we will be to robots what dogs are to humans, /
"我在想像一个时刻 / 那时机器人之于我们就像狗之于人类，/
and I'm rooting for the machines." /
而我支持机器。"/

——**克劳德·艾尔伍德·香农**，《通信的数学原理》作者

"Some people worry / that artificial intelligence will make us feel inferior, / but then, /
"一些人担心 / 人工智能会让我们感觉到自卑，/ 但那时，/
anybody in his right mind should have an inferiority complex / every time he looks at a flower." /
所有人都应该有自卑感 / 每当他看到花的时候。"/

——**艾伦·凯**，美国科学家

"The sad thing about artificial intelligence is / that it lacks artifice / and therefore intelligence." /
"人工智能最悲哀之处在于 / 缺乏技巧 / 也就因此缺少智能。"/

——**让·鲍德里亚**，哲学家

"Intelligence is the art of good guesswork." /
"智能是推测的艺术。"/

——**贺拉斯·巴兹尔·巴洛**，*The Oxford Companion to the Mind* 作者

"Artificial Intelligence is a two-edged sword. / On the one hand, / it allows us /
"人工智能是把双刃剑。/ 一方面，/ 它允许我们 /
to create intelligent artifacts with human-like perception and cognition. / On the other hand, /
创造具备堪比人类感觉和认识的智能产品。/ 另一方面，/
it accelerates / people's heavy dependence / on artifacts." /
它加速了 / 人类的过度依赖 / 对人工产品的。"

——**马克斯·布拉莫**，*Artificial Intelligence: An International Perspective* 作者

"Today's AI is about new ways / of connecting people to computers, /
"今日的 AI 是关于新方式的，/ 连接人类和计算机、/
people to knowledge, / people to the physical world, / and people to people." /
人类和知识、/ 人类和物理世界 / 以及人与人的。"/

——**帕特里克·温斯顿**，麻省理工学院教授，摘自 MIT 人工智能实验室简报

第三部分

大数据

“大数据”是当下的一大热词，它在众多领域都功不可没。本部分整理了相关案例和研发人员（特别是求职时）需要深度解析大数据的原因。

大数据，高收益
Big Data, Big Pay

For anyone evaluating a big data career, here are some facts that will help make your decision easier.

- According to an Accenture Survey, 56% of executives believe they do not have the talent to analyze the data collected from a big data analytics implementation [1].
- McKinsey Global Institute predicts the United States alone will face a shortage of 140,000 to 190,000 analytics experts by 2018 [2].
- A report by the Tech Partnership and SAS anticipates the UK will create approximately 56,000 big data jobs per year until 2020 [3].

Yes, it's clear that demand for big data expertise is on the rise. The industry is already experiencing a shortage of talent, which is expected to become a major barrier to growth.

What does it all mean to current and future IT professionals? It means there are great job opportunities in the field of big data and analytics. At Indeed.com, as of March 2015, the search term "big data" yields over 35,000 jobs, with Hadoop ranking 7th in top job trends in the IT industry.

Let's say you have made a decision to explore potential job opportunities in big data. Where to begin? Start with understanding the most common job titles used by companies seeking expertise in the field. It will help you broaden your perspective on the industry.

RCR Wireless suggests 12 common job titles listed on top job sites [4]:

> Data Scientist, Data Engineer, Big Data Engineer, Machine Learning Scientist, Business Analytics Specialist, Data Visualization Developer, Business Intelligence (BI) Engineer, BI Solutions Architect, BI Specialist, Analytics Manager, Machine Learning Engineer, Statistician

Among the titles above, data scientist is the hottest job in today's big data market. A Harvard Business Review article even called data scientist "the sexiest job of the 21st century."[6] Currently Indeed.com lists more than 17,000 "data scientist" jobs with an average salary of $117,000 [5].

So how to become a data scientist? Plenty of information is available on the web, but the industry leaders' job listings are the best resources for identifying the technical skills and backgrounds required to enter the field. Go to major tech companies' websites and run a search. IBM's career page, for example, yields 101 job listings for the search term "data scientist," each describing required and preferred skills in detail. Go ahead and check it out for yourself!

出处

1. Big Data Analytics Widens Talent Gap, http://goo.gl/fWFQtL
2. Big data: The next frontier for competition, http://goo.gl/MD3R4
3. Investment needed to meet UK demand for big data skills and analytics, http://goo.gl/6hRqmu
4. Top 12 Big Data & Analytics Job Titles, http://goo.gl/7N4DtG
5. Data Scientist: The Sexiest Job of the 21st Century, https://goo.gl/VJhbHu

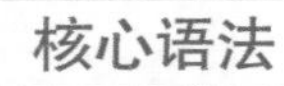

核心语法

熟悉 # 关系代名词。# 关系代名词连接两个句子，兼具 # 连词和 # 代名词的作用。例文将介绍多种关系代名词句。# 关系代名词有章可循，理解原理后参照例句学习将事半功倍。

- For **anyone** ([**who** is] evaluating a big data career), here are some facts (**that** will help make your decision easier). 为大家，/评估大数据职业的/这有/一些案例/可帮助大家轻松做决定。
- They do not have people to analyze **the data** ([**that** is] collected from a system). 他们没有人力/分析数据/从系统中收集的。
- **The industry is experiencing a shortage of talent**, **which** is expected to become a major barrier to growth. 业界正经历人力匮乏期，这将成为预想的/主要障碍/发展的。

熟悉 # 分词从句。# 分词从句是缩短句子的一种方式，表推测的部分可省略。例文将介绍因主语不同而不可省略的句子。以下两个例句中，如果分词前没有主语，则表示前后两个句子主语一致。

- The page yields many job listings, **each describing** required skills. 此页面展示了很多招聘列表，每个都列出了必备的工作技能。
  = The page yields many job listings, **and each job listing describes** required skills.
  =页面展示了很多招聘列表，并且每个招聘列表都列出了必备的工作技能。
- **The page** yields many job listings, **describing** required skills. 此页面展示了很多招聘列表，列出了必备的工作技能。
  = The page yields many job listings, **and the page describes** required skills. =此页面展示了很多招聘列表，并且该页面列出了必备的工作技能。

熟悉表附属关系的 #with 分词从句〈with+ 宾语 + 宾补〉。宾语和宾补是主谓关系，意为“当 ~ 的时候，在 ~ 的状态下”。

- The search term “big data” yields over 35,000 jobs, **with** Hadoop **ranking** 7th in top job trends. 关键字big data创造出35 000多个工作机会/Hadoop位列第七/在就业趋势中。

熟悉常将 #that 引导的从句用作 # 宾语的动词。that 常省略。

- 56% of executives **believe** (that) ~ 56%的管理人员相信/~
- The institute **predicts** (that) ~ 机构预言/~
- A report **anticipates** (that) ~ 报告预计/~

单词&短语

evaluate 评估
fact 事实
decision 决定
According to ~ 依据 ~
executive 管理人员
implementation 贯彻，实现
predict 预计
face 直面
shortage 不足
analytics expert 分析家
anticipate 预测
approximately 大约
demand 需要
expertise 专业知识
on the rise 增加趋势
barrier to ~ 阻碍 ~
growth 增长
opportunity 机会
yield 生产
rank 等级，排名
make a decision 决定
explore 探险
potential 潜在的
broaden 拓宽
perspective 观点，视角
suggest 建议
job title 职务
article 报道
currently 现在
average salary 平均工资
plenty of ~ 充足的 ~
resource 资源

require 要求	preferred 更喜欢的
search term 检索词	in detail 详细地
describe 说明，描述	check out 调查，确认

根据提示完成句子

_____ ___ / you have ____ _ __________ /
例如/ 你已经决定/

to explore _________ ___ _______________ / in big data. /
探寻潜在的工作机会/ 在大数据领域。/

_____ __ _____? / _____ ____ understanding the most common ___ ______ /
该从哪里着手呢？/ 首先从理解最基本的职责开始/

used by companies seeking expertise / __ ___ _____. /
被公司用于查找专业知识/ 在专业领域。

It will help / you _______ ____ ______________ / __ ___ __________. /
这将帮助你/ 拓宽视野/ 在业界。

So ___ __ ______ a data scientist? / ______ __ information is available /
那该如何成为数据科学家呢？/ 有大量信息可供使用/

on the web, / but the industry leaders' job listings are the best resources /
在网络上，/ 但业界先驱的职位列表是最优资源/

for _____________ the ___________ ______ and _____________ /
为了掌握最专业的技术和背景/

required to _____ ___ _____. /
以进入这个领域。/

Go to major tech companies' websites / and run a search. /
访问主要技术公司的网站/ 去查询吧。

思考题

讨论

1 请列举数据专家需具备的知识。（提示：http://goo.gl/4PpL）

2 请列举与大数据相关的职业类型。（提示：http://goo.gl/Q8NVPb）

课堂总结

1 学习#关系代名词和#关系代名词省略的概念和例句。

2 学习#分词从句构成方法和#with分词从句的概念和例句。

3 学习常将#that从句用作#宾语的动词和例句。

翻译

Big Data, Big Pay 大数据，高收益

For anyone evaluating a big data career, /
为评估大数据职业的每个人，/
here are / some facts / that will help make your decision easier. /
这里 / 有些案例 / 可帮助大家轻松做决定。/

- According to an Accenture Survey, / 56% of executives believe / they do not have the talent /
  据埃森哲调查，/56%的管理层认为/他们没有才能/
  to analyze the data collected / from a big data analytics implementation [1]. /
  分析收集到的数据/从大数据分析中。/
- McKinsey Global Institute predicts /
  麦肯锡全球研究所预计/
  the United States alone will face a shortage of 140,000 to 190,000 analytics experts /
  美国将面临缺少14万~19万分析家的状况/
  by 2018 [2]. /
  到2018年。/
- A report by the Tech Partnership and SAS anticipates /
  Tech Partnership和SAS报告预计/
  the UK will create approximately 56,000 big data jobs / per year / until 2020 [3]. /
  英国将创造约56 000个大数据职位/每年/直到2020年。/

Yes, it's clear / that demand for big data expertise is on the rise. /
是的，/ 很明显 / 对大数据专业知识的需求在不断攀升。/
The industry is already experiencing a shortage of talent, /
业界正面临人才枯竭，/
which is expected / to become a major barrier / to growth. /
这将是 / 巨大障碍 / 发展的。/

What does it all mean / to current and future IT professionals? /
这将意味着什么呢？ / 对现在和未来的 IT 专家 /
It means / there are great job opportunities / in the field of big data and analytics. /
这意味着 / 将有大量工作机会 / 在大数据和分析领域。/
At Indeed.com, / as of March 2015, / the search term "big data" yields over 35,000 jobs, /
在 Indeed.com，/ 以 2015 年 3 月为准，/ 关键字 big data 产生了 35 000 多个职位，/
with Hadoop ranking 7th / in top job trends / in the IT industry. /
Hadoop 位列第七 / 在就业趋势中 /IT 业界。/

Let's say / you have made a decision / to explore potential job opportunities / in big data. /
例如 / 你已经决定 / 探寻潜在的工作机会 / 在大数据领域。/

Where to begin? / Start with understanding the most common job titles /
该从何处着手呢？ / 首先从理解最基本的职责开始 /
used by companies seeking expertise / in the field. /
被公司用于查找专业知识 / 在专业领域。/
It will help / you broaden your perspective / on the industry. /
这将帮助你 / 拓宽业界视野。/

RCR Wireless suggests 12 common job titles / listed on top job sites [4]: /
RCR Wireless 建议将 12 种常规职务 / 列举在工作网站顶端：/
Data Scientist, Data Engineer, Big Data Engineer, Machine Learning Scientist,
数据科学家 / 数据工程师 / 大数据工程师 / 机器学习科学家 /
Business Analytics Specialist, Data Visualization Developer, Business Intelligence (BI) Engineer,
商业分析专家 / 数据可视化研发 / 商业信息（ BI ）工程师 /
BI Solutions Architect, BI Specialist, Analytics Manager, Machine Learning Engineer, Statistician /
商业信息解决方案架构师 / 商业信息专家 / 分析经理 / 机器学习工程师 / 统计分析师 /

Among the titles above, / data scientist is the hottest job / in today's big data market. /
以上众多职务中，/ 数据科学家是最具人气的职位 / 在当今大数据市场。/
A Harvard Business Review article even called data scientist / "the sexiest job of the 21st century."[6] /
《哈佛商业评论》的一篇报道甚至称数据科学家为 / "21 世纪最吸引人的工作"。/
Currently / Indeed.com lists more than 17,000 "data scientist" jobs / with an average salary of $117,000 [5]. /
现在 /Indeed.com 列举了 17 000 余个 "数据科学家" 职位 / 平均年薪为 117 000 美元。/

So how to become a data scientist? / Plenty of information is available / on the web, /
那么如何才能成为数据科学家呢？ / 有大量信息可供使用 / 在网络上，/
but the industry leaders' job listings are the best resources /
但是业界先驱的职位列表是最优资源 /
for identifying the technical skills and backgrounds / required to enter the field. /
为了掌握最专业的技术和背景 / 以进入这个领域。/
Go to major tech companies' websites / and run a search. /
访问主要技术公司的网站 / 去查询吧。/
IBM's career page, / for example, / yields 101 job listings / for the search term "data scientist," /
IBM 的求职页面中，/ 例如，/ 展示了 101 个职位列表 / 通过搜索 data scientist，/
each describing / required and preferred skills / in detail. / Go ahead and check it out / for yourself! /
各自描述了 / 必备的技能 / 详细地。/ 去调查吧 / 你亲自！ /

研发人员的招聘秘诀：以实力取胜

Recruiting Top Developers

Lately, a growing number of companies are leveraging big data for human resources and recruiting [1], and Gild is a company that offers technology specifically designed for recruiting software developers. The company used its own software to hire a Rails programmer in a most unconventional way. Here's the story [1, 2].

Jade Dominguez didn't quite fit the typical profile for a programmer. He didn't go to college. He taught himself programming because he needed a website for a custom T-shirts company he had started. But he was obsessed with programming and contributed a lot to GitHub. He was living off credit card debt when he received an email offering an interview opportunity as a programmer at a San Francisco startup called Gild.

Luca Bonmassar, co-founder and chief executive at Gild, had discovered Jade through Gild's proprietary software called Gild Source. Gild Source evaluates code contributed to various open source platforms and developer communities — such as GitHub, Bitbucket, Google Code, LinkedIn, Twitter, Stack Overflow, and Facebook — looking for the right talent. Gild says that it goes "where developers hang out" — where they interact and display their code and knowledge — and scores millions of developers to offer recruiters deeper insight into a candidate's potential for a position. Gild Source provides access to over 6 million developer profiles. In addition, Gild gathers each candidate's social media activity to help companies determine culture fit.

When Luca and his team ran a search for a Rails programmer using Gild Source, Jade was at the top of the list. He had no college education or work experience, but Gild Source gave him a phenomenal score based on his work on GitHub. He had built a solid reputation on GitHub, and his code for Jekyll-Bootstrap had been reused by over 1,000 developers. His code displayed expertise in Rails and JavaScript. His blog postings and tweets were opinionated, which was something the company wanted. Luca and the other developers on the team were impressed.

The story says that Jade wore a vibrant green hoodie to the interview. The interview went well, some pointed questions were asked and answered, and the company offered him a job on the spot. He accepted a position with an annual salary of around $115,000. Imagine a guy with no college degree and no professional background getting a job at a hot Silicon Valley startup!

出 处

1. Big Data and Recruiting: The Resources You Need, http://goo.gl/cplMvX
2. How Big Data Is Playing Recruiter for Specialized Workers, http://goo.gl/ccvrr
3. How Gild, Inc. Used Big Data to Recruit a Hard-to-Find Developer, https://goo.gl/3dPjGv

核心语法

熟悉 # 过去完成时态〈had+ 过去分词〉。表示过去某一时间或动作以前已经发生或完成了的动作对过去的某一点造成的某种影响或结果，指在另一个过去行动之前就已经完成了的事件。它表示动作发生的时间是“过去的过去”，侧重事情的结果。

- Luca sent Jade an email. Luca **had discovered** Jade. 卢卡给杰德发了邮件。卢克已经发现了杰德。
- He needed a website for a company [**that**] he **had started**. 他需要一个网站/给公司/他已经运营的。
- The software gave him a phenomenal score. He **had built** a solid reputation on GitHub. 这个软件给了他一个惊人的分数。他已经在GitHub大有名气。（=所以在软件给他分数之时，他已经大有名气。）

熟悉 # 关系代名词的省略。# 宾格关系代名词和 # 主格关系代名词 +be 动词可以省略。省略后不影响原意。

- Gild offers **technology** [**which** is] specifically designed for recruiting developers. Gild提供技术/它是特别设计的/为招聘研发。
- He needed a website for **a company** [**that**] he had started. 他需要一个网站/为一家公司/他已经运营的。

熟悉 # 关系副词。# 关系副词连接两个句子，兼具 # 连词和 # 副词的作用。原理同 # 关系代名词一致，但 # 先行词或者 # 关系副词的省略条件不一致。以下例句中可使用任一表达。

- The software goes **the place**. + Developers hang out **in the place**. 软件去了这个地方。 + 研发娱乐的/场所。

  = The software goes **the place where** developers hang out. 软件去了这个地方/研发娱乐的。

  = The software goes **the place** developers hang out.

  = The software goes **where** developers hang out.

  ▶ the place和where同时存在时，可省略其一。

熟悉 # 分词从句和 # 关系代名词的省略。从句子结构上看，以下例句适用于两种解释。找出分词的主语，根据上下文理解即可。

- His team ran a search **using** the software. 他的团队进行了检索/用软件。

  ▶ 如果视为分词从句，那么是句子的主语“他的团队”使用软件。

- His team ran a search [which was] **using** the software. 他的团队进行了检索/使用软件。

  ▶ 如果视为关系代名词从句，那么是先行词“检索”使用软件。

- He received an email **offering** an interview opportunity. 他收到了邮件/提供了面试机会。

  ▶ 如果视为分词从句，那么是句子的主语“他”提供。

- He received an email [which was] **offering** an interview opportunity. 他收到了邮件/提供面试机会的。

  ▶ 如果视为关系代名词从句，那么是先行词“邮件”提供。

单词&短语

a number of 大量的
growing 增加的
leverage 使用
specifically 特别地
hire 雇佣
unconventional 与众不同的，不依惯例的
quite 十分，相当
fit 适合
typical profile 典型经历
college 大学
teach oneself ~ 自学 ~
be obsessed 痴迷于 ~
contribute A to B 将 A 献给 B
live off ~ 以 ~ 为生
receive 接受
offer 提供
opportunity 机会
co-founder 共同创始人
chief executive 首席执行官
discover 发现

proprietary 专有的
evaluate 评价
look for 寻找
hang out 闲逛
insight 洞察力
potential 潜力

技术术语

Jekyll–Bootstrap 是 Jekyll 的增强版，采用静态文件方式管理，不需要数据库即可支持一个独立博客站点，github-pages 平台普遍采用该款软件。

根据提示完成句子

When Luca and his team ran a search / for a Rails programmer /
当卢卡和他的团队搜索/ 为了寻找Rails程序员/

using Gild Source, / Jade was __ ___ ___ __ the list. /
使用Gild Source，/ 杰德处于列表顶端。/

He had __ _______ _________ or ____ ___________, /
他没有任何大学教育和工作经历，/

but Gild Source ____ him / a ___________ score / _____ __ his work /
但Gild Source给了他/ 一个惊人的分数/ 基于他的工作/

on GitHub. /
在GitHub上。/

He had _____ a _____ ___________ / on GitHub, /
他获得一致好评/在GitHub，/

and his code for Jekyll-Bootstrap / had been _______ /
并且他的Jekyll–Bootstrap代码 / 已被使用/

__ ____ 1,000 developers. / His code displayed __________ /
由1000余名研发人员。/ 他的代码很专业/

in Rails and JavaScript. / His blog postings and tweets were ___________, /
在Rails和JavaScript方面。/ 他的博文和推文很有自己的想法，/

which was something the company wanted. /

这恰恰符合公司的要求。/

Luca and the other developers / __ ___ ____ / were impressed. /

给卡卡和其他研发人员/ 团队里的/ 留下了深刻印象。

思考题

解答题

1 （理解）以下哪种SNS不适合用于评估研发人员？

ⓐ LinkedIn ⓑ GitHub

ⓒ Stack Overflow ⓓ Kakao Talk

2 （论述）在GitHub或LinkedIn SlideShare上传自己的作品集对履历管理有何帮助？

讨论

1 会有公司使用大数据招聘研发人员吗？（提示：http://goo.gl/qf7c8j）

2 根据Gild 提供的宣传视频（https://goo.gl/7SwaER），探讨Gild和现存HR行业的差异。

课堂总结

1 学习#现在完成时和#过去完成时的概念和例句。

2 学习#关系代名词和#关系副词的概念和例句。

3 学习#分词从句的概念和例句。

答案

1 d.（Kakao Talk多用于个人社交） 2 个人擅长的领域获得专业认可的可能性变大。

翻译

Recruiting Top Developers

研发人员的招聘秘诀：以实力取胜

Lately, / a growing number of companies / are leveraging big data /
最近 / 越来越多的公司 / 在使用大数据 /
for human resources and recruiting [1], /
用于人力资源和招聘，/
and Gild is a company / that offers technology / specifically designed /
Gild 是一家公司 / 它提供技术 / 特别设计 /
for recruiting software developers. /
为招聘软件研发人员。/
The company used its own software / to hire a Rails programmer /
这个公司使用它自己的软件，/ 招聘 Rails 程序员 /
in a most unconventional way. / Here's the story [1, 2]. /
以最独特的方式。/ 详情如下。/

Jade Dominguez didn't quite fit the typical profile / for a programmer. /
杰德・多明戈斯并不完全符合要求 / 作为程序员。/
He didn't go to college. / He taught himself programming /
他没上过大学。/ 自学编程 /
because he needed a website / for a custom T-shirts company / he had started. /
因为他需要一个网站 / 为一个定制 T 恤的公司 / 他运营的。/
But he was obsessed / with programming / and contributed a lot / to GitHub. /
但他执着于 / 编程 / 并且做出了很多贡献 / 对 GitHub。/
He was living off credit card debt / when he received an email / offering an interview opportunity /
他靠信用卡借债为生 / 当他收到邮件 / 提供面试机会的 /
as a programmer / at a San Francisco startup / called Gild. /
作为一个程序员 / 在旧金山创业公司 / 名为 Gild。/
Luca Bonmassar, / co-founder and chief executive / at Gild, / had discovered Jade /
卢卡・邦马萨，/ 共同创始人和首席执行官 /Gild 的，/ 发现了杰德 /
through Gild's proprietary software / called Gild Source. / Gild Source evaluates code /
借助 Gild 的专属软件 / 名为 Gild Source 的。/Gild Source 评估代码 /
contributed to various open source platforms and developer communities /
捐赠给各种公开平台和研发团体的 /
– such as GitHub, Bitbucket, Google Code, LinkedIn, Twitter, Stack Overflow, and Facebook – /
——例如 GitHub、Bitbucket、Google Code、LinkedIn、Twitter、Stack Overflow 和 Facebook——/
looking for the right talent. / Gild says / that it goes "where developers hang out" /
以寻找合适的人才。/Gild 称 / 这些显示出"研发人员消耗时间的场所"/

– where they interact and display their code and knowledge – /
——研发人员相互联系以及展示代码和知识的场所——/
and scores millions of developers / to offer recruiters deeper insight /
并且为数百万名研发人员评分 / 以方便人事负责人更深入观察 /
into a candidate's potential / for a position. / Gild Source provides access /
应聘者的潜力 / 对于岗位的。/Gild Source 能获取 /
to over 6 million developer profiles. /
600 万研发人员的简历。/
In addition, / Gild gathers each candidate's social media activity /
此外，/Gild 收集了各位应聘者的社交媒体活动 /
to help / companies determine culture fit. /
以帮助 / 公司决定文化匹配度。/

When Luca and his team ran a search / for a Rails programmer / using Gild Source,
当卢卡和他的团队搜索 / 为了寻找 Rails 程序员 / 使用 Gild Source/
Jade was at the top of the list. / He had no college education or work experience, /
杰德处于列表顶端。/ 他没有任何大学教育或工作经历，/
but Gild Source gave him / a phenomenal score / based on his work / on GitHub. /
但 Gild Source 给了他 / 一个惊人的分数 / 基于他的工作 / 在 GitHub。/
He had built a solid reputation / on GitHub, / and his code for Jekyll-Bootstrap /
他获得一致好评 / 在 GitHub，/ 并且他的 Jekyll–Bootstrap 代码 /
had been reused / by over 1,000 developers. / His code displayed expertise /
已被使用 / 由 1000 余名研发人员。/ 他的代码很专业 /
in Rails and JavaScript. / His blog postings and tweets were opinionated, /
在 Rails 和 JavaScript 方面。/ 他的博文和推文很有自己的想法，/
which was something the company wanted. /
这恰恰符合公司的要求。/
Luca and the other developers / on the team / were impressed. /
给卢卡和其他研发人员 / 团队里的 / 留下了深刻印象。/

The story says / that Jade wore a vibrant green hoodie / to the interview. /
据说 / 杰德穿了绿色连帽衫 / 面试时。/
The interview went well, / some pointed questions were asked and answered, /
面试很顺利，/ 他被问了很多尖锐的问题，/
and the company offered him a job / on the spot. / He accepted a position /
公司聘用了他 / 当场。/ 他接受了这份工作 /
with an annual salary of around $115,000. /
以约 115 000 美元的年薪。/
Imagine a guy / with no college degree and no professional background /
设想一个人 / 没有大学学历和专业背景 /
getting a job / at a hot Silicon Valley startup! /
得到了一份工作 / 在竞争激烈的硅谷创业公司！ /

大数据之大

How BIG Is Big Data?

Everyone talks about "Big Data." It's the hottest trend in the industry now. Some argue that Big Data will change the future of business. Various reports and research findings support the claim.

According to Wikibon's Big Data Vendor Revenue and Market Forecast report [1], the Big Data market will grow from $10.2 billion in 2013 to $50 billion by 2017. Accenture's Industrial Insights Report for 2015 [2] indicates that 73 percent of companies are already investing more than 20 percent of their overall technology budget on Big Data analytics.

Well, then it must be important, right? But what is Big Data? How big is it really?

What is Big Data?

Most people think that the term Big Data means an extremely large amount of data. That's partially true. It used to mean just a massive volume of data, but now it is an umbrella term to encompass methods and technologies to gather, analyze, and interpret such massive data. A recent White House report describes Big Data as our "growing technological ability to capture, aggregate, and process an ever-greater volume, velocity, and variety of data."[3]

How big is Big Data?

Let's take a specific example: Twitter. Statistics show that there are, on average, 350,000 tweets per minute [4]. Assume that tweets contain an average of 15 words each. Then it's 5.25 million words per minute. If someone prints those words, since the standard number of words per page is 250 words, they would fill 21,000 pages. If you stack those pages and measure the height of that stack,

- In 1 minute, it would measure 2.1 meters.
- In 2 hours, it would be as tall as the 63 Building in Seoul.
- In 3 days, it would be taller than Mt. Everest.
- In 5 weeks, it would reach outer space.

Twitter is just one example and tweet text is just one form of data. Google alone processes over 40,000 search queries every second, 3.5 billion searches per day, and 1.2 trillion searches per year [LIVESTATS]. YouTube gets about 100 hours of new videos every minute and users watch over 6 billion hours of videos every month [5]. Facebook has 1.3 billion monthly active users and handles 3 million messages every 20 minutes [6].

That is how BIG, Big Data is. In fact, BIG may not be a big enough word.

出 处

1. Big Data Vendor Revenue and Market Forecast 2013-2017, http://goo.gl/tZ71Ni
2. How the Industrial Internet is Changing the Competitive Landscape of Industries, http://goo.gl/sB4hoh
3. BIG DATA: SEIZING OPPORTUNITIES, PRESERVING VALUES, http://goo.gl/WjGeaz
4. Internet Live Stats, http://www.internetlivestats.com/
5. YouTube Statistics, https://goo.gl/92m41l
6. Facebook Statistics, http://goo.gl/ZoLju

核心语法

熟悉 # 句式 3〈主语 + 动词 + 宾语〉。用 # 名词和 #that 从句作 # 宾语。#that 从句作 # 宾语时，that 可省略。

- Some argue **that** Big Data will change the future of business. 一些人认为/大数据将改变商业的未来。
- The report indicates **that** companies are already investing money. 报告显示/公司已经开始注资。
- Assume **that** tweets contain an average of 15 words each. 假设/每条推文平均有15个字。

熟悉 # 不定式 to 的 # 形容词用法。接在名词后修饰名词，意为“~的，将~的”，表 # 限定。

- It is a term **to encompass all technologies**. 这个术语包含所有技术。
- It is a term to encompass all technologies **to analyze such massive data**. 这个术语包含所有分析大数据的技术。
- It is technological ability **to process an ever-greater volume of data**. 这是一种处理空前大数据量的技术能力。

熟悉 # 并列句。以下例句依然采用的是 # 不定式 to 作形容词的例句，最后一项前均加 # 并列连词 and。

- It is a term to encompass **methods and technologies**. 这个术语包含所有技术和手段。
- It is a term to encompass methods and technologies to **gather, analyze, and interpret** such massive data. 这个术语包含收集、分析和诠释大数据的众多手段和技术。
- It is technological ability to **capture, aggregate, and process** ever-greater volume of data. 这是一种收集、分析和处理庞大数据的技术能力。

熟悉 # 比较句。例文将介绍常用句型。

- It would be **as** tall **as** the 63 Building. 它将与63大厦同高。
- It would be tall**er than** Mt. Everest. 它将高于珠穆朗玛峰。

单词&短语

argue that ~ 主张 ~
support a claim 支持主张
according to ~ 依据 ~
revenue 利润，销量
forecast report 预测报告
indicate that ~ 预示 ~
invest 投资
overall budget 总预算
must be ~ 必定是 ~
extremely large amount 极多量
partially 部分地
used to V 过去常常做 ~
massive volume 大量的
umbrella term 宏观用语
encompass 围绕，包围
gather 收集
analyze 分析
interpret 解析
describe A as B 将 A 描述为 B
aggregate 综合
velocity 速度
variety 多样性
specific example 具体案例
statistics 统计学
assume (that) ~ 假设 ~
stack 堆积
measure the height of ~ 测量 ~ 的高度
as tall as ~ 高达 ~
reach 到达
search query 搜索查询
in fact 事实上

技术术语

Big Data 大数据指无法在一定时间范围内用常规软件工具进行捕捉、管理和处理的数据集合，是需要新处理模式才能具有更强决策力、洞察力和流程优化能力的海量信息资产，具有高增长率和多样化的性质。

根据提示完成句子

Everyone _____ ____ "Big Data." / It's ___ _______ ______ /
所有人都在谈论"大数据"。/ 这是个热门趋势/

__ ___ __________ / now. / Some _____ /
在业界/ 最近。/ 一些人认为/

that Big Data ____ ______ ___ ______ __ ___________. /
大数据将改变商业的未来。/

Various ______ ___ _________ _________ ________ ___ ______. /
众多报告和研究结果支持以上观点。/

What is Big Data? / 何为大数据?

Most people think / that ___ ____ ___ ____ _____
很多人认为/ "大数据"意味着/

an _________ _____ ______ __ ____./
极其大的数据量。/

That's _________ ____. / It used to ____ ____ _ ________ ______ __ ____,
这只说对了一部分。/ 它过去仅指大量数据，/

but now it is an _________ ____ /
但现在是一个宏观术语/

to __________ _______ ___ ___________ / to ______, ________,
针对庞大数据的/ 收集、分析和解释的/

and _________ ____ _______ ____. /
包含众多技术和手段。/

A recent _____ _____ ______ __________ Big Data /
最新的白宫报告称大数据/

__ ___ "_______ _________ _______ / to _______, __________,
是我们"日益增长的技术能力/ 比任何时候都大/

and ________ __ ____–______ ______, ________, and ______ of data."[3] /
高速获取、收集和处理各种数据的。"/

思考题

解答题

1 （理解）以下哪种数据不适用大数据技术？

ⓐ SNS 数据　　ⓑ 人类基因信息

ⓒ 搜索引擎索引信息　　ⓓ 飞机设计图

2 （理解）以下哪一项没有正确阐述大数据？

ⓐ 随着电脑越来越强大，大数据的重要性逐渐减弱。

ⓑ 3V（volume、velocity、variety）很重要

ⓒ SNS 相关企业用于处理数据。

ⓓ 目前最热门的技术趋势之一。

3 （论述）为什么不能仅用“非常大的数据”解释大数据？

4 （论述）收集、分析和解析数据的过程中，哪个阶段要求同时具备对业界的洞察力和科学的处理方式？

讨论

1 思考保存大数据的技术。现存技术有哪些局限？有哪些解决途径？

2 国内有可以处理“大”数据的数据处理公司或者团体吗？

课堂总结

1 学习#that从句作#名词的例句。

2 学习#不定式to用作#形容词的例句。

3 学习使用#原级、#比较级、#最高级的#比较句。

答案

1 d.（飞机设计图中确实有很多零部件，但存在局限性。） 2 a.（为处理更多数据，技术在发展，市场也在不断扩展。） 3 不仅要考虑数量、数据的速度，也要考虑多样性。 4 分析阶段（除了洞察力，也需要相关领域的专业知识和处理众多信息的数学以及统计学背景知识。）

翻译

How BIG Is Big Data?

大数据之大

Everyone talks about "Big Data." / It's the hottest trend / in the industry / now. /
所有人都在谈论 "大数据"。/ 这是个热门趋势 / 在业界 / 最近。/
Some argue / that Big Data will change the future of business. /
一些人认为 / 大数据将改变商业的未来。/
Various reports and research findings support the claim. /
众多报告和研究结果支持以上观点。/

According to Wikibon's Big Data Vendor Revenue and Market Forecast report [1], /
据 Wikibon 的大数据销售商收入和市场预测报告，/
the Big Data market will grow / from $10.2 billion in 2013 / to $50 billion by 2017. /
大数据市场将增长 / 从 2013 年的 102 亿美元 / 到 2017 年的 500 亿美元。/
Accenture's Industrial Insights Report for 2015 [2] indicates /
Accenture 2015 年产业展望报告显示 /
that 73 percent of companies are already investing /
73% 的公司已经投资 /
more than 20 percent of their overall technology budget / on Big Data analytics.
超过整体技术预算的 20%/ 在大数据分析。/

Well, / then it must be important, / right? / But what is Big Data? / How big is it really?
那么，/ 它一定很重要，/ 不是吗？ / 但何为大数据？ / 大数据究竟有多大？ /

What is Big Data? / 何为大数据?

Most people think / that the term Big Data means an extremely large amount of data. /
很多人认为 / "大数据" 术语意味着极其大的数据量。/
That's partially true. / It used to mean just a massive volume of data, / but now it is an umbrella term /
这只说对了一部分。/ 它过去仅指大量数据，/ 但现在是一个宏观术语 /
to encompass methods and technologies / to gather, analyze, and interpret such massive data. /
包含众多收集、分析和解释庞大数据的技术和手段。/
A recent White House report describes Big Data / as our "growing technological ability /
最新的白宫报告称大数据 / 是我们 "日益增长的技术能力 /
to capture, aggregate, and process an ever-greater volume, velocity, and variety of data."[3] /
以获取、收集和处理具有超大容量和速度的各种数据。" /

How big is Big Data? / 大数据究竟有多大?

Let's take a specific example: / Twitter. / Statistics show that / there are, / on average, /
让我们看一个具体案例：/Twitter。/ 统计显示 / 平均 /
350,000 tweets per minute [4]. / Assume that / tweets contain an average of 15 words each. /
每分钟有 35 万条推文。/ 假设 / 每条推文平均有 15 个字。/
Then it's 5.25 million words per minute. / If someone prints those words, /
那么每分钟就有 525 万个字被发出。/ 如果将这些字都打印出来，/
since the standard number of words per page is 250 words, /
标准纸张每张有 250 个字，/
they would fill 21,000 pages. / If you stack those pages / and measure the height of that stack,
则需要打印 21 000 张。/ 如果将这些纸堆起来 / 并测量高度，/

In 1 minute, / it would measure 2.1 meters. /
1分钟内，/将高达2.1米。/
In 2 hours, it would be as tall as the 63 Building in Seoul. /
2小时内，将与63大厦同高。/
In 3 days, it would be taller than Mt. Everest. /
3天内，将高于珠穆朗玛峰。/
In 5 weeks, it would reach outer space. /
5周内，将高耸入云。/

Twitter is just one example / and tweet text is just one form of data. /
Twitter 仅是其中一个案例 / 并且推文仅是一种数据形式。/
Google alone processes / over 40,000 search queries / every second, /
谷歌单独处理 / 超过 4 万条搜索查询 / 每秒，/
3.5 billion searches per day, / and 1.2 trillion searches per year [LIVESTATS]. /
每天 35 亿次查询，/ 每年 12 000 亿次查询。/
YouTube gets / about 100 hours of new videos / every minute / and users watch /
YouTube 播放 / 约 100 小时的全新视频 / 每分钟 / 并且用户观看 /
over 6 billion hours of videos / every month [5]. / Facebook has 1.3 billion monthly active users /
超过 60 小时的视频 / 每月。/Facebook 有 13 亿月活跃用户 /
and handles 3 million messages / every 20 minutes [6].
处理 300 万条信息 / 每 20 分钟。/

That is how BIG, / Big Data is. / In fact, / BIG may not be a big enough word. /
这就是庞大之处 / 大数据的。/ 事实上，/“大”可能并不能充分表达。/

IBM 让城市更智慧

IBM Helps Cities Get Smarter

IBM is a clear leader in the Big Data and Analytics market. The company ranked No. 1 for two consecutive years, according to Wikibon's Big Data Vendor Revenue and Market Forecast report [1]. But IBM isn't just a market leader. The company shares its big data expertise with cities around the world. One good example is IBM's Smarter Cities Challenge [2].

The Smarter Cities Challenge is IBM's largest philanthropic program. Every year, cities around the globe apply for the challenge. The selected cities receive a grant from IBM.

The grant, however, is not in the form of financial aid. IBM does not give money to the cities. Instead, the company provides free consultation. IBM sends a team of five or six experts to each city for three weeks. During that period, the IBM team analyzes the city's data and meets with stakeholders. At the end of the three weeks, the team delivers strategic recommendations.

Over the past four years, IBM has helped more than 115 cities worldwide. On average, each city received about $400,000 worth of consulting services. Two Korean cities, Cheongju and Jeju, are among them. Winning cities implemented the IBM team's recommendations to improve the lives of their citizens [3]. Glasgow, United Kingdom, is a great example [3].

In Glasgow, the issue was fuel poverty. More and more families couldn't keep warm at reasonable cost. The city wanted to bring affordable warmth to its citizens. An IBM team of 5 staff met over 30 organizations and nearly 100 individuals from the City Council and other agencies. The team analyzed data and identified inefficient energy use as a major cause of fuel poverty. That was a problem for the whole community, not just for the poor families. At the end of three weeks, the IBM team presented a list of 60 recommendations. The recommendations focused on improving energy literacy throughout the city. The team believed that citizens would use less energy if they understood their energy usage better.

Glasgow is working hard to achieve its goal of becoming the most energy literate city in Europe. In 2013, the city won a $40 million grant from the UK's Technology Strategy Board to become Britain's first smart city [4].

Originally IBM was going to run the Smarter Cities Challenge only for three years, but the program has been extended because of its huge success.

出 处

1. Big Data Vendor Revenue and Market Forecast 2013-2017, http://goo.gl/tZ71Ni

2. About the Smarter Cities Challenge, http://goo.gl/CmwH1N

3. Challenges and cities, http://goo.gl/ebvWM

4. Glasgow will be UK's first 'smart city', http://goo.gl/WqAXD

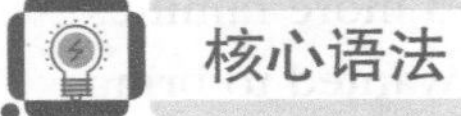

核心语法

熟悉 # 句式 2〈主语 + 动词 + 补语（名词、形容词）〉。使用 #be 动词和 # 延续性动词 keep。#be 动词意为“是”，# 延续性动词 keep 意为“使保持 ~”或者“继续 ~”。

- IBM **is** a clear market leader. IBM毫无疑问是市场的领头羊。
- The issue **was** fuel poverty. 问题是燃料不足。
- Families couldn't **keep** warm. 家庭无法保持供暖。

熟悉 # 句式 3〈主语 + 动词 + 宾语〉。# 名词和 # 不定式 to 作 # 宾语。# 不定式 to 作 # 名词时，意为“做 ~”。want 后可以接 # 不定式 to 作 # 宾语。

- The selected cities receive a grant. 被选中的城市可获得补助金。
- The company provides free consultation. 公司提供免费咨询。
- The city wanted **to bring** affordable warmth to its citizens. 城市希望 / 给市民提供价格低廉的取暖设备。

熟悉 # 不定式 to 作 # 副词的 # 副词用法。可表目的（为了 ~）、原因（因为 ~）、条件（如果 ~）、结果（做了 ~）等。

- The cities implemented the IBM team's recommendations **to improve** the lives of their citizens. 城市落实了IBM的提议/为了改善市民生活。
- The city is working hard **to achieve** its goal. 城市在努力工作/以完成目标。
- The city won a grant **to become** Britain's first smart city. 城市赢得了补助/成为英国首个智慧城市。

熟悉 # 句式 4〈主语 + 动词 + 间接宾语 + 直接宾语〉，与 # 句式 3 意义相同。

- IBM does not give money **to** the cities. IBM不向城市提供资金。
  = IBM does not give cities money.
- IBM sends a team of five or six experts **to** each city. IBM向各个城市派遣了5~6人的专家组。
  = IBM sends each city a team of five or six experts.
- The city wanted to bring affordable warmth **to** its citizens. 城市希望/给市民提供价格低廉的取暖设备。
  = The city wanted to bring its citizens affordable warmth.

单词&短语

rank 占据（等级，排名）
consecutive 连续的
revenue 利润
forecast 预告
share A with B 与 B 分享 A
philanthropic 慈善的，博爱的
apply for 申请
grant 补助
financial aid 财政补助
free consultation 免费咨询
analyze data 数据分析
stakeholder 利益相关者
strategic recommendation 战略建议
on average 平均地
implement 贯彻，落实
improve 改善
fuel poverty 燃料匮乏
at reasonable cost 以合理的花费
affordable 低廉的
nearly 几乎
individual 个人
agency 团体，机构
identify A as B 将 A 视为 B
inefficient energy use 能源利用率低下
a major cause 主要原因
present 发表
focus on ~ 致力于 ~
energy literacy 能源素养
originally 原来，本来
be going to V 将要做 ~
extend 延伸，延长
because of ~ 因为 ~

技术术语

smart city 智慧城市就是运用信息和通信技术手段感测、分析、整合城市运行核心系统的各项关键信息，从而对包括民生、环保、公

共安全、城市服务、工商业活动在内的各种需求做出智能响应。其实质是利用先进的信息技术，实现城市智慧式管理和运行，进而为城市中的人创造更美好的生活，促进城市的和谐、可持续发展。

根据提示完成句子

______, / however, / is not in the ____ __ _____ ___. /
补助，/ 但，/ 并不是财政援助。

IBM does not ____ _____ to the cities. /
IBM并不向城市提供资金。/

Instead, / the company ________ ____ __________. /
取而代之，/ 它提供免费咨询。/

IBM sends ____ __ five or six ______ / to each city / ___ ____ ____. /
IBM派遣5~6位专家/ 前往各大城市/ 为期3周。/

_______ that period, / the IBM team ________ the city's ____ ___ _____ /
在此期间，/ IBM团队分析城市的数据和需求/

with ___________. / __ ___ ___ __ the three weeks, /
与利益相关者。/ 3周后，/

the team ______ __________ _____________. /
专家组提交战略提案。/

__ ___ ___ __ three weeks, /
3周后，/

the IBM team _________ a ____ of 60 ________________. /
IBM专家组提交60份提案列表。/

The ________________ _______ __ _________ ________ ________ /
提案关注改善能源素养/

___________ the city. / The team believed /
整个城市的。/ 专家组相信/

that ________ would ___ ____ ______ /
市民将减少使用能源/

if they _________ their ______ _____ better. /
如果他们能更准确地了解自己的能源使用量。

思考题

解答题

1 （理解）以下哪一项与例文内容一致？

ⓐ IBM 向城市注资。 ⓑ IBM 在城市构建计算机设施。

ⓒ IBM 免费给城市咨询。 ⓓ IBM 在城市构建基础设施。

2 （理解）以下与智慧城市相关的内容中，哪一项与格拉斯哥无关？

ⓐ 燃料匮乏现象

ⓑ 仅穷人会面临寒冷灾害

ⓒ 咨询过程中 IBM 提供 60 种提案

ⓓ 提案的重点在于改善能源利用能力

3 （论述）能源素养和燃料匮乏之间存在何种关系？

4 （论述）IBM使用大数据分析技术从事慈善业务的核心是什么？

讨论

1 除可利用大数据技术节约能源外，请列举可以推动城市改革的其他方案。

2 请列举利用城市现有基础设施可收集到的数据类型。个人信息保护相关数据收集过程中会遇到何种阻碍？

课堂总结

1 学习#句式2中使用的动词和例句。

2 学习可将#不定式to用作#宾语的动词和例句。

3 学习#不定式to的#副词用法。

答案

1 c. 2 b. 3 只有解决能源素养问题，才能有效使用能源，避免能源匮乏。 4 通过分析城市大数据，找出实现城市战略发展的方案，并给出建议的免费咨询业务。

翻译

IBM Helps / Cities Get Smarter
IBM让城市更智慧

IBM is a clear leader / in the Big Data and Analytics market. / The company ranked No. 1 /
IBM 毫无疑问是市场的领头羊 / 在大数据和分析市场中。/IBM 位列第一 /
for two consecutive years, /
连续 2 年，/
according to Wikibon's Big Data Vendor Revenue and Market Forecast report [1]. /
据 Wikibon 的大数据业界收益和市场预测报告分析。/
But IBM isn't just a market leader. / The company shares its big data expertise /
但 IBM 并不仅仅是市场领头羊。/ 这家公司将其大数据经验共享 /
with cities around the world. / One good example is IBM's Smarter Cities Challenge [2]. /
给全球各大城市。/ 其中最好的例证是 IBM 的 "智慧城市挑战"。/

The Smarter Cities Challenge is IBM's largest philanthropic program. /
"智慧城市挑战" 是 IBM 最大的慈善项目。/
Every year, / cities around the globe / apply for the challenge. /
每年，/ 来自全球的各个城市 / 都申请挑战。/
The selected cities receive a grant from IBM. /
被选中的城市将获得 IBM 的补助。/

The grant, / however, / is not in the form of financial aid. / IBM does not give money to the cities. /
补助，/ 但，/ 并不是财政援助。/IBM 并不向城市提供资金。/
Instead, / the company provides free consultation. / IBM sends a team of five or six experts /
取而代之，/ 它提供免费咨询。/IBM 派遣 5~6 位专家 /
to each city / for three weeks. /
前往各大城市 / 为期 3 周。/
During that period, / the IBM team analyzes the city's data and meets /
在此期间，/IBM 团队分析城市的数据和需求 /
with stakeholders. / At the end of the three weeks, /
与利益相关者。/3 周后，/
the team delivers strategic recommendations. /
专家组提交战略提案。/

Over the past four years, / IBM has helped more than 115 cities worldwide. /
过去 4 年内，/IBM 已在全球范围内援助超过 115 个城市 /
On average, / each city received about $400,000 worth / of consulting services. /
平均，/ 每个城市获得价值 400 000 美元的 / 咨询服务。/
Two Korean cities, / Cheongju and Jeju, / are among them. /
两个韩国城市，/ 清州和济州 / 就位列其中。/

Winning cities implemented the IBM team's recommendations /
获胜的城市执行 IBM 专家组的提案 /
to improve the lives of their citizens [3]. /
以改善市民生活。/
Glasgow, / United Kingdom, / is a great example [3]. /
格拉斯哥，/ 英国的 / 是很好的案例。/

In Glasgow, / the issue was fuel poverty. / More and more families couldn't keep warm /
在格拉斯哥，/问题是燃料匮乏。/越来越多的家庭无法保持供暖/
at reasonable cost. / The city wanted to bring affordable warmth / to its citizens. /
以合理的价格。/城市希望提供价格低廉的取暖设备/给市民。/
An IBM team of 5 staff / met over 30 organizations and nearly 100 individuals /
一个5人IBM专家组/会见了30多家机构和100多名个人/
from the City Council and other agencies. / The team analyzed data /
来自城市委员会和其他机构。/专家组分析数据/
and identified inefficient energy use / as a major cause of fuel poverty. /
发现能源利用率低下/是燃料匮乏的主要原因。/
That was a problem / for the whole community, / not just for the poor families. /
这是个大问题/对整个共同体而言/并不仅仅针对贫困家庭。/
At the end of three weeks, / the IBM team presented a list of 60 recommendations. /
3周后，/IBM团队提交了60份提案列表。/
The recommendations focused on improving energy literacy / throughout the city. /
提案关注改善能源素养/整个城市的。/
The team believed / that citizens would use less energy /
专家组相信/市民将减少能源使用/
if they understood their energy usage better. /
如果他们能更准确地了解自己的能源使用量。/

Glasgow is working hard / to achieve its goal /
格拉斯哥在努力/完成目标/
of becoming the most energy literate city in Europe. /
成为欧洲能源素养最好的城市。/
In 2013, / the city won a $40 million grant from the UK's Technology Strategy Board /
2013年，/这个城市从英国技术战略委员会获得了4000万美元/
to become Britain's first smart city [4]. /
成为英国首个智慧城市。/

Originally / IBM was going to run the Smarter Cities Challenge / only for three years, /
最初 /IBM 计划运营 “智慧城市挑战” / 仅 3 年，/
but the program has been extended / because of its huge success. /
但这个计划延长 / 因为取得了巨大成功。/

天气预报公司跻身广告界翘楚

How the Weather Company Became an Advertising Powerhouse

The Weather Company is a leader in weather forecasting. Through the Weather Channel and Weather.com, the company provides national and local weather forecasts. It serves hundreds of thousands of people, and provides the information for thousands of mobile weather apps.

The company has been using big data to improve its prediction abilities for the last decade. "Weather is the original big data application," says Bryson Koehler, CIO at the Weather Company. "When mainframes first came about, one of the first applications was a weather forecasting model."[1] The company's current system captures 2.2 million current-weather-condition data points from around the globe four times per hour. The company's new consolidated platform, which is under development, will capture 2.25 billion weather data points 15 times per hour [2].

But now the Weather Company is using its big data more than just to forecast the weather. It analyzes the behavior patterns of its digital and mobile users in 3 million locations worldwide to predict how the weather affects people's shopping patterns. With this analytics data, the company has become an advertising powerhouse [3].

One example is a Pantene shampoo ad campaign that was on the Weather Company's mobile app [4]. Depending on the city and its weather, the app displayed different ads that recommended different kinds of shampoo. For example, if humidity was on the rise, the app would suggest a shampoo that prevents frizz. Pantene immediately saw spikes in sales from this campaign and signed on for more ads [5].

In an interview with the *Atlantic Monthly*, the company's ad tech guru Vikram Somaya gave a good overview on how the Weather Company help retailers with weather data: "Retail is an easy example. We have a retailer who may have a couple hundred or even a couple thousand stores across the U.S. We take data from each of those locations for each of their products, then we look at the information over time. We are looking to see what products start jumping off shelves when the dew point is X, the temperature is Y, and the rainfall is Z. What we give them is essentially: Here are the 15 products you should be selling right now."[4] It's no surprise that more than half of the Weather Company's ad revenue comes from its digital operations [3].

出处

1. Big Data Reshapes Weather Channel Predictions, http://goo.gl/XDlrUk
2. Up Next: The Weather Channel Forecasts the Business Value of Big Data, http://goo.gl/9jf2JT
3. The World's Top 10 Most Innovative Companies in Big Data, http://goo.gl/zXlMfm
4. Cloudy With a Chance of Beer, http://goo.gl/kOpXKD
5. Weather Co. exec: Leverage seasonal data to deliver relevant ads, http://goo.gl/J8FEFg

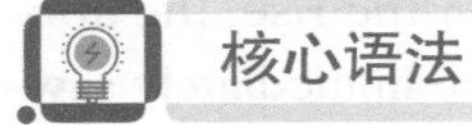

核心语法

熟悉 # 现在完成时态〈have/has+ 过去分词〉。指过去发生的事情对现在造成的影响。

- The company **has been using** big data. 这家公司一直在使用大数据。（=所以现在也在使用）
- The company **has become** an advertising powerhouse. 这家公司已经成为广告界翘楚。（=所以现在是广告界翘楚）

熟悉 # 关系代名词。# 关系代名词连接两个句子，兼具 # 连词和 # 代名词的作用。例文将介绍多种关系代名词句。# 关系代名词有章可循，理解原理并结合例句学习将事半功倍。

- We have **a retailer**. + **It** may have a couple hundred stores. 我们有零售商。+ 它可能有数百家店。

  = We have **a retailer** (**who** may have a couple hundred stores). =我们有零售商/可能有数百家店。
- The app displayed different **ads** (**that** recommended different kinds of shampoo). 这个应用展示不同广告/推荐不同种类洗发水的。
- **The thing** [**that**] we give them is this. 那个东西/我们给他们的/是这个。

  = **What** we give them is this. =我们给他们的东西是这个。
- Here are **the 15 products** ([**that**] you should be selling). 这15种产品/你应该出售。
- The company's **new platform**, **which** is under development, will capture weather data points. 公司的新平台，目前正在发展中的，将捕捉天气数据点。

熟悉 # 间接疑问句。# 间接疑问句指直接引语是疑问句，变为间接引语。含义较简单，但写作时需注意语序。

- The company predicts **how** the weather affects people's shopping patterns. 公司预测/天气如何影响人的购物模式。
- She gave a good overview on **how** the Weather Company help retailers with weather data. 她给出了一个关于~的不错的概述（=简单说明）/气象公司如何用天气数据帮助零售商。
- We are looking to see **what products** start jumping off shelves. 我们正期待看到/何种产品开始畅销。

单词&短语

weather forecast 天气预报
provide 提供
national 全国的
local 当地的
hundreds of ~ 成百的 ~
thousands of ~ 成千的 ~
hundreds of thousands of 几十万，无数
improve 改善
prediction ability 预测能力
decade 10 年
original 原来的
mainframe 大型机
come about 出现，发生
capture 捕捉
data point 数据点
consolidated platform 综合平台
under development 研发中的
analyze 分析
behavior pattern 行为模式
predict 预计
affect 影响
analytics data 分析数据
advertising powerhouse 广告（界）一把手
depending on ~ 根据 ~
recommend 推荐
humidity 湿度
on the rise 在上涨的
suggest 建议
prevent 防止
frizz 使卷曲
spike 长钉，尖状物
overview 概要
retailer 零售商
product 产品
over time 随着时间的流逝
dew point 露点

技术术语

Supercomputer 超级计算机因为计算量庞大，世界各国气象机构纷纷引进。美国 NOAA-NWS（National Oceanic and Atmospheric Administration-National Weather Service，海洋暨大气总署－国家气象局）将超级计算机的处理能力由 90 teraflops 升级到 213 teraflops，2015 年升级到 1950 teraflops。详情参考 http://goo.gl/KfvXnn。

根据提示完成句子

In an ________ ____ the *Atlantic Monthly*, /
在《大西洋月刊》的一篇采访中，/

the company's ad tech guru Vikram Somaya gave a good overview /
公司的广告技术负责人维克拉姆·索马亚给出了一个很好的总结/

on how the Weather Company help _______ /
关于气象公司如何帮助零售商/

with weather data: / "______is an easy example. / We have a _______ /
获取天气数据:/ "零售是个简单的例证。我们有零售商/

who may have a couple hundred or even a couple thousand stores /
可能有上百家店铺/

across the U.S. / We take data / from each of those locations /
遍及美国/ 我们采集数据/ 从各地/

for each of their products, / then we look at the information / ____ ____. /
分品类，/ 然后我们查看信息/ 随着时间的流逝。/

We are _______ __ ___ / what products _____ _______ ___ ________ /
我们期待看到/何种产品开始畅销/

when the ___ _____ is X, / the ___________ is Y, /
当露点是X，/ 气温是Y，/

and the ________ is Z. / What we give them / is essentially: /
降雨是Z时。/ 我们给他们的/

Here are the 15 products / you should be selling right now."[4] /
基本如下：/这15款产品/你应该立刻出售。" /

It's __ ________ /
毫无疑问/

that more than half of the Weather Company's __ ________ /
气象公司超过一半的广告收入/

comes / from its digital operations [3]. /
来自于/数字运营。

思考题

解答题

1 （理解）以下哪项技术与天气预报无关？

ⓐ 天气应用程序　　ⓑ 大型机

ⓒ 实时掌握动物移动轨迹　　ⓓ 大数据分析

2 （论述）请简述天气预报与大数据的关系。

3 （论述）请列举用天气预报创造商业利润的案例。

讨论

1 请用手机下载知名天气应用（如“雅虎天气”），实际使用之后，你建议增加何种功能？

2 气象厅提供的天气预报涵盖各个领域，被广泛使用。如果要提高私营机构天气预报的附加值，该向谁额外提供何种内容？

课堂总结

1 学习#现在完成时态的概念和例句。

2 学习#关系代名词的概念和例句。

3 学习#疑问句和#间接疑问句的概念和例句。

答案

1 c. 2 以每小时 4次、220万个数据标准收集资料，分析覆盖300万个地区的数字和移动用户行为模式。 3 零售店根据天气更换库存商品，针对不同天气消费者的购买模式进行广告宣传。

翻译

How the Weather Company Became an Advertising Powerhouse 天气预报公司跻身广告界翘楚

The Weather Company is a leader / in weather forecasting. /
气象公司是领头羊 / 天气预报领域的。/
Through the Weather Channel and Weather.com, /
通过天气频道和 Weather.com，/
the company provides national and local weather forecasts. /
公司提供全国和各地天气预报。/
It serves hundreds of thousands of people, / and provides the information /
它服务于成百上千的人群，/ 并且提供信息 /
for thousands of mobile weather apps. /
为数以千计的移动端天气应用。/

The company has been using big data / to improve its prediction abilities /
公司一直在使用大数据 / 提高自己的预测能力 /
for the last decade. / "Weather is the original big data application," / says Bryson Koehler,
在过去 10 年。/ "天气是最原始的大数据应用，" / 布莱森・凯勒说，/
CIO at the Weather Company. / "When mainframes first came about, /
气象公司首席信息官。/ "大型机刚出现的时候，/
one of the first applications was a weather forecasting model."[1] /
首个应用之一就是天气预报模型。" /
The company's current system captures 2.2 million current-weather-condition data points /
公司现有系统捕捉 220 万当前天气状况数据点 /
from around the globe / four times per hour. / The company's new consolidated platform, /
在全球 / 每小时 4 次。/ 公司的最新综合平台，/
which is under development, / will capture 2.25 billion weather data points /
目前正在研发中，/ 将能捕捉 22.5 亿天气数据点。/
15 times per hour [2]. /
以每小时 15 次的速度。/

But now the Weather Company is using its big data / more than just to forecast the weather. /
但是，现在气象公司正在使用大数据 / 不仅为了天气预报。/
It analyzes the behavior patterns / of its digital and mobile users / in 3 million locations /
它分析行为模式 / 数字和移动用户的 / 来自 300 万个场所 /
worldwide / to predict / how the weather affects people's shopping patterns. /
全球的 / 目的在于预测 / 天气如何影响人们的购物模式。/

With this analytics data, / the company has become an advertising powerhouse [3]. /
通过这个分析数据，/ 公司已经成为广告界翘楚。/
One example is a Pantene shampoo ad campaign / that was on the Weather Company's mobile app [4]. /
案例之一是潘婷洗发水广告营销案 / 它在气象公司的移动应用上进行推广。/
Depending on the city and its weather, / tthe app displayed different ads /
根据城市和天气，/ 应用展示不同广告 /
that recommended different kinds of shampoo. / For example, /
推荐不同种类的洗发水。/ 例如，/
if humidity was on the rise, / the app would suggest a shampoo / that prevents frizz. /
如果湿度增加，/ 应用会建议使用洗发水 / 防止卷发的。/
Pantene immediately saw spikes / in sales / from this campaign /
潘婷立刻看到激增 / 销量的 / 通过这个广告系列 /
and signed on / for more ads [5]. /
于是签约 / 以投放更多广告。/

In an interview with the Atlantic Monthly, /
在《大西洋月刊》的一篇采访中 /
the company's ad tech guru Vikram Somaya gave a good overview /
公司的广告技术负责人维克拉姆·索马亚给出了一个很好的总结 /
on how the Weather Company help retailers / with weather data: /
关于气象公司如何帮助零售商 / 获取天气数据：/
"Retail is an easy example. /
"零售是个简单的例证。/
We have a retailer / who may have a couple hundred or even a couple thousand stores /
我们有零售商 / 可能有上百家店铺 /
across the U.S. / We take data / from each of those locations / for each of their products, /
遍及美国 / 我们采集数据 / 从各地 / 分品类 /
then we look at the information / over time. / We are looking to see /
然后我们查看信息 / 随着时间的流逝。/ 我们期待看到 /
what products start jumping off shelves /
何种产品开始畅销 /
when the dew point is X, / the temperature is Y, / and the rainfall is Z. / What we give them /
当露点是 X，气温是 Y，降雨是 Z 时。/ 我们给他们的 /
is essentially: / Here are the 15 products / you should be selling right now."[4] /
基本如下：/ 这 15 款产品 / 你应该立刻出售。"/
It's no surprise / that more than half of the Weather Company's ad revenue / comes /
毫无疑问 / 气象公司超过一半的广告收入 / 来自 /
from its digital operations [3]. /
数字运营。/

经典语录之大数据篇
Notable Quotes on Big Data

Let's hear what IT industry leaders say about Big Data & Analytics.

"Big data is not about the data" — **Gary King**, professor at Harvard University

"Data really powers everything that we do." — **Jeff Weiner**, chief executive of LinkedIn

"You can have data without information, but you cannot have information without data." — **Daniel Keys Moran**, computer programmer and science fiction author

"There were 5 exabytes of information created between the dawn of civilization through 2003, but that much information is now created every 2 days." — **Eric Schmidt**, former CEO of Google

"I keep saying that the sexy job in the next 10 years will be statisticians, and I'm not kidding" — **Hal Varian**, chief economist at Google

"Information is the oil of the 21st century, and analytics is the combustion engine." — **Peter Sondergaard**, analyst at Gartner Research

"Big data is at the foundation of all the megatrends that are happening today, from social to mobile to cloud to gaming."

— **Chris Lynch**, CEO of Vertica Systems

"Without big data, you are blind and deaf in the middle of a freeway"

— **Geoffrey Moore**, management consultant and theorist

"Data is the new science. Big Data holds the answers."

— **Pat Gelsinger**, CEO of VMware, Inc.

"Data are becoming the new raw material of business."

— **Chris Lynch**, former President and chief executive officer of Vertica Systems.

"The goal is to turn data into information, and information into insight."

— **Carly Fiorina**, former executive, president, and chair of Hewlett-Packard Co.

"If we have data, let's look at data. If all we have are opinions, let's go with mine."

— **Jim Barksdale**, former Netscape CEO

"If you torture the data long enough, it will confess to anything."

— **Darrell Huff**, author of *How to Lie With Statistics*

"Data that is loved tends to survive."

— **Kurt Bollacker**, Data Scientist, Freebase/Infochimps

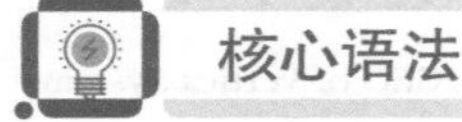

核心语法

学习 # 关系代名词 that。# 关系代名词可连接两个句子，兼具 # 连词和 # 代名词的作用。修饰 # 关系代名词从句的名词为 # 先行词。以下例句中，everything、megatrends、data 即为 # 先行词。

- Data powers **everything**. + We do **it**. 数据给所有的事情以动力。+ 我们做的。
  = Data powers **everything that** we do. =数据给所有的事情以动力/我们做的。
- Big data is at the foundation of all the **megatrends**. + **They** are happening today. 大数据是所有大趋势的基石。+ 它们正发生在今天。
  = Big data is at the foundation of all the **megatrends** (**that** are happening today). =大数据是所有大趋势的基石/它们正发生在今天。
- **Data** tends to survive. + **It** is loved. 数据多会留存下来。+ 它被人喜爱。
  = **Data** (**that** is loved) tends to survive. =数据/被人喜爱的/多会留存下来。

学习 # 关系代名词 what。# 关系代名词 what 相当于 the thing(s) that。包含 # 先行词，意为"～做～"。

- Let's hear **things**. + IT industry leaders say **them** about Big Data. 让我们听取说法。+ IT业界领军人物关于大数据的说法。
  = Let's hear **things** (**that** IT industry leaders say things about Big Data). =让我们听取说法/IT业界领军人物关于大数据的说法。
  = Let's hear **what** IT industry leaders say about Big Data. =让我们听取/IT 业界领军人物关于大数据的说法。

熟悉 # 关系代名词的省略。# 宾格关系代名词、# 主格关系代名词 +be 动词省略后不影响原意。# 宾格在原句中作 # 宾语，# 主格作 # 主语。

- Data powers **everything**. + We do **it**. 数据给所有的事情以动力。+ 我们做的。
  = Data powers **everything** (**that** we do). =数据给所有的事情以动力/我们做的。
  = Data powers everything [**that**] we do.
  ▶ 宾格that可以省略，关系代名词we do后没有宾语，可以省略。
- There was a lot of **information**. + **They** was created last year. 这有很多信息。+ 它们于去年被创造出来。
  = There was a lot of **information** (**that** was created last year). =这有很多信息/它们于去年被创造出来。
  = There was a lot of **information** [**that** was] created last year.
  ▶ 主格+be动词that was省略后不影响原意。

单词&短语

analytics 分析
power 力量，提供动力
dawn 凌晨
civilization 文明
keep ~ing 持续做 ~
statistician 统计专家
combustion engine 内燃机
foundation 基础
blind 盲人
deaf 听障人士
in the middle of ~ 在 ~ 中间
freeway 高速公路
science 科学
hold 捉住
raw material 原材料
turn A into B 将 A 变为 B
insight 洞察力
opinion 意见
torture 拷问
confess to 承认，坦白
tend to V ~ 倾向于做 ~
survive 幸存

技术术语

Data science 数据科学指从数据中挖掘知识的科学。包括数学、统计学、信息技术（信号处理、概率模型、机器学习、统计学习、计算机编程、数据工程、模式识别与学习、可视化、预测分析、不确定性建模、数据仓库、数据压缩、高性能计算机）等领域出现的技术和理论。数据科学与大数据有密切联系。

根据提示完成句子

"You can have ____ / without __________, /
"没有信息/ 可以获取数据，/

but you cannot have __________ / without ____."
但没有数据/ 不能获取信息。"

— **Daniel Keys Moran**, computer programmer and science fiction author

"__________ is the ___ of the 21st century, /
"如果说信息是21世纪的机油，/

and ________ is the _________ ______."
那么分析就是内燃机。"

— **Peter Sondergaard**, analyst at Gartner Research

"If you _______ the data / ____ ______, /
"如果你充分拷问数据/ 长时间地/

it will _______ __ _________."
它会承认所有事情。"

— **Darrell Huff**, author of *How to Lie With Statistics*

"Without big data, /

"没有大数据，/

you are _____ and ____ / in the ______ of a _______."

就如同一个聋哑人/ 站在高速公路中间。"

— **Geoffrey Moore**, management consultant and theorist

"Data is the ___ _______. /

"数据是一门新的科学。/

Big Data _____ the ________."

答案尽在大数据手中。"

— **Pat Gelsinger**, CEO of VMware, Inc.

思考题

讨论

1 当今，数据较匮乏的领域也在强调"大数据"。在这种形势下，请再次思考大数据的本质。

课堂总结

1 学习#关系代名词的概念和种类。

2 学习#关系代名词引导非限定性定语从句。

3 学习#关系代名词的省略。

翻译

Notable Quotes / on Big Data

经典语录之大数据篇

Let's hear / what IT industry leaders say about Big Data & Analytics. /
让我们听取 /IT 业界领军人物关于大数据的经典说法。/

"Big data is not about the data."
"大数据超越了数据。"/

——**加里·金**，哈佛大学教授

"Data really powers everything / that we do."
"数据给所有事情以动力 / 我们做的。"/

——**杰夫·维纳**，领英首席执行官

"You can have data / without information, / but you cannot have information / without data."
"没有信息 / 可以获取数据，/ 但没有数据 / 不能获取信息。"/

——**丹尼尔·吉斯·莫兰**，计算机程序员、科学小说作者

"There were 5 exabytes of information created / between the dawn of civilization through 2003, /
"5 EB 信息被创造 / 在 2003 年文明之初，/
but that much information is now created / every 2 days."
而现在等量信息被创造 / 每 2 天。"/

——**埃里克·施密特**，前谷歌首席执行官

"I keep saying / that the sexy job / in the next 10 years / will be statisticians, / and I'm not kidding."
"我一直在说 / 最性感的职业 / 在未来 10 年 / 将是统计专家 / 我不是在开玩笑。"/

——**哈尔·瓦里安**，谷歌首席经济学家

"Information is the oil of the 21st century, / and analytics is the combustion engine."
"如果说信息是 21 世纪的机油，/ 那么分析就是内燃机。"/

——**彼得·桑德加德**，Gartner 分析中心分析师

"Big data is at the foundation of all the megatrends / that are happening today, /
"大数据是所有大趋势的基石 / 发生在今天的，/
from social to mobile to cloud to gaming."
从社交媒体到移动设备、云、游戏。"/

——**克里斯·林奇**，前 Vertica Systems 首席执行官

“Without big data, / you are blind and deaf / in the middle of a freeway.”

“没有大数据，/ 就如同一个聋哑人站在高速公路中间。”/

——**杰弗里·摩尔**，战略与创新咨询专家

“Data is the new science. / Big Data holds the answers.”

“数据是一门新的科学。/ 答案尽在大数据手中。”/

——**帕特·基尔辛格**，VMware 首席执行官

“Data are becoming the new raw material of business.”

“数据正在成为全新的商业原材料。”/

——**克里斯·林奇**，前 Vertica Systems 首席执行官

“The goal is to turn data into information, / and information into insight.”

“目的是将数据转换为信息，/ 将信息转化为洞察力。”/

——**卡莉·菲奥莉娜**，前惠普公司首席执行官

“If we have data, / let’s look at data. / If all we have are opinions, / let’s go with mine.”

“如果我们有数据，/ 那我们看数据。/ 如果我们有各自的想法，/ 那我们遵循我的想法。”/

——**吉姆·巴克斯代尔**，前 Netscape 首席执行官

“If you torture the data / long enough, / it will confess to anything.”

“如果你充分拷问数据 / 长时间地，/ 它会承认所有事情。”/

——**达莱尔·哈夫**，*How to Lie With Statistics* 作者

“Data / that is loved / tends to survive.”

“数据 / 被喜爱的 / 多会留存下来。”/

——**柯特·博莱克**，Freebase/Infochimps 数据科学家

第四部分

物联网

继大数据之后，物联网横扫市场。当然，也有人会认为IoT只是M2M的变形。但不能否认，物联网已经悄然走入人们的日常生活。本部分将着重探讨几个具体案例。

日益智能的路灯

Street Lights Are Getting Smarter

It is well known that street lighting provides a number of benefits. Street lighting improves road visibility for motorists and pedestrians, so it reduces traffic accidents. It is also known to reduce the severity of injuries by a factor of three [1]. Moreover, street lighting reduces crime and fear of crime and promotes the nighttime economy.

Street lighting, however, has its own drawbacks. It consumes a significant amount of electricity and requires continuous maintenance and repair, all of which put pressure on a city's budget. Street lighting is also a major source of light pollution. Light pollution, or excessive artificial lighting, is found to have negative effects on both human and wildlife health [2].

That is why cities around the globe are adopting Internet of Things technology for their streets. Major cities have already started to replace their existing street lights with more efficient light-emitting diodes (LEDs). New York plans to replace 250,000 lights by 2017 and expects an annual savings of $14 million [3]. Seoul will replace all public lighting with LEDs by 2018 and expand the project to the private sector by 2030 [4].

In addition to energy-efficient LEDs, cities are also connecting street lights and sensors through a network to enable smart street lighting. According to Telematics Wireless [5], smart street lighting systems offer the following benefits:

- They reduce maintenance costs. The system detects and sends alerts for lamp burnouts and other malfunctions in real time, which enables the maintenance crew to take timely actions. No need for routine maintenance checks.
- They allow dimming-intensity control. The system can automatically dim the street lights or turn them on/off based on various conditions set by operators or triggered by sensors. For example, the lights are dimmed when there's no traffic. This achieves major energy savings as well as a reduction in light pollution.
- They extend the lifespan of street lights. LEDs have a longer life than conventional street lamps. Smart dimming extends their life even longer, which reduces maintenance costs.
- They offer real-time reporting and control. The system collects real-time data such as voltage, current, power factor, and energy consumption to help operators manage the grid in a smarter way.

This is how your city is becoming smarter. The street lamp you passed by last night may be more intelligent than you think!

出 处

1. Street Lighting and Road Safety , http://goo.gl/KdV2Ak
2. Improved street lighting can reduce crime, http://goo.gl/rw5QK6
3. How LEDs Are Going To Change The Way We Look At Cities, http://goo.gl/eOX84g
4. More LEDs to light Seoul City streets, http://goo.gl/y0k87E
5. What Does the Word "Smart" Stand for in Smart Street Lighting? http://goo.gl/8Hlf4s

核心语法

熟悉 # 关系代名词引导非限定性定语从句。用于补充说明前句。# 关系代名词前加逗号（,）可以理解为〈连词 + 代名词〉。例文将介绍代指前句的 # 关系代名词 which。

- It requires continuous maintenance and repair, **all of which** put pressure on a city's budget. 它需要持续的维护和修理，/所有的一切都加重了负担/城市预算的。
- The system detects lamp burnouts in real time, **which** enables the maintenance crew to take timely actions. 系统实时监测路灯故障，/以方便维修人员能及时采取措施。
- Smart dimming extends their life even longer, **which** reduces maintenance costs 智能调光灯延长了寿命，/这减少了维修费用。

熟悉 # 关系副词。# 关系副词可连接两个句子，兼具 # 连词和 # 副词的作用。原理类似 # 关系代名词，但 # 先行词和 # 关系副词的省略条件不同。以下例句中，可使用任一表达。

- That is **the reason**. + Cities are adopting this technology **for the reason**. 这就是原因。+ 城市采取这项技术/因为这个原因。

  = That is **the reason** (**for which** cities are adopting this technology). =这就是原因/城市采取这项技术的。

  = That is **the reason why** cities are adopting this technology.

  = That's **the reason** cities are adopting this technology.

  = That is **why** cities are adopting this technology.

  ▶ the reason和 why同时存在，可省略其一。
- This is **the way**. + Your city is becoming smarter **in the way**. 就是这种方式。+ 你的城市正变得越来越智能/以这种方式。

  = This is **the way** (**in which** your city is becoming smarter). =就是这种方式/使你的

城市正变得越来越智能的。

= This is **the way** your city is becoming smarter.

= This is **how** your city is becoming smarter.

▶ the way和how不能同时使用，二者只能选其一。

熟悉 # 被动语态。that 引导的宾语从句变成被动语态时，要使用 # 形式主语 it。使用 # 不定式 to 也可缩短句子。理解原理后无需死记硬背 be known to 和 befound to 之类的表达。

- People know that street lighting provides many benefits. 人们知道/路灯有很多好处。
  = **It is known that** street lighting provides many benefits. =据悉/路灯有很多好处。
  = Street lighting **is known to** provide many benefits. =路灯被知道/有很多好处。
- People found that light pollution has negative effect on human health. 人们发现/光污染有不良影响/对人体健康。
  = **It is found that** light pollution has negative effect on human health. =据发现/光污染有不良影响/对人体健康。
  = Light pollution **is found to** have negative effect on human health. =光污染被发现/有不良影响/对人体健康。

单词&短语

provide 提供
street lighting 路灯
a number of 大量的
benefit 益处
improve 改善
visibility 可视性
motorist 司机
pedestrian 行人
reduce 减少，缩减
traffic accident 交通事故
severity 严重程度
injury 伤害，受伤
by a factor of ~ 是 ~ 倍、~ 倍于
moreover 并且
fear of crime 畏惧犯罪
promote 促进
nighttime economy 夜间经济
drawback 缺点
consume 消费
an amount of ~ 大量的
electricity 电力
require 要求
continuous 持续的
maintenance 维护，管理
repair 修理
alert 警报

技术术语

Smart grid 智能电网指在电力的生产、运输、消费过程中引入计算机技术，使供应商和消费者以最优状态消费电力的智能型电网系统。消费者和电力公司实时交换信息，消费者在电力优惠的时间段用电，供应商实时把握用电量，进行弹性调整。

根据提示完成句子

It is ____ _____ / that ______ _______ provides a number of benefits. /
众所周知/ 路灯有很多优点。/

Street lighting improves ____ __________ /
路灯提高道路可视度/

for _________ and ___________, / so it reduces ______ _________. /
为司机和行人，/ 所以减少了交通事故。/

It is also _____ / to ______ the ________ of ________ /
同时/ 减少受伤的严重程度/

__ _ _______ __ three. /
到原来的1/3。

Moreover, / street lighting reduces crime and fear of crime /
并且，/ 路灯减少犯罪和对犯罪的恐惧/

and ________ the _________ _______. /
以及促进夜间经济。/

Street lighting, / however, / has its ___ _________. /
路灯，/ 但是，/ 有自身的不足。/

It _________ a significant amount of electricity /
它消耗大量电力/

and _______ __________ ___________ and ______, /
并且需要持续的维护和修理，/

all of which put _______ / on a city's budget. /
所有一切加重了负担/ 给城市预算。/

Street lighting is also a _____ ______ / of _____ _________. /
路灯也是主要的来源/ 光污染的。/

_____ _________, / or excessive artificial lighting, / is found /
光污染，/ 或过度的人工照明，/ 被发现/

to have ________ _______ / on both human and ________ health.
有不良的影响/对人类和野生动物的健康。

思考题

解答题

1 （理解）以下哪一项不属于智能路灯的优点？

ⓐ 减少维护费用 ⓑ 可调节亮度

ⓒ 费用低廉 ⓓ 实时调控

2 （论述）请列举降低光污染的方法。

讨论

1 请简述将路灯换为LED智能灯最大的障碍以及解决途径。

课堂总结

1 学习#关系代名词引导非限定性定语从句的概念和例句。

2 学习#关系副词的概念和例句。

3 学习#被动语态的概念和例句。

答案

1 c.（相比一般灯泡，LED灯泡的价格较高。） 2 根据车辆通行频率和周边环境（比如月光）调节亮度以减少光污染。

翻译

Street Lights Are Getting Smarter
日益智能的路灯

It is well known / that street lighting provides a number of benefits. /
众所周知 / 路灯有很多优点。/
Street lighting improves road visibility / for motorists and pedestrians, /
路灯提高道路可视度 / 为司机和行人，/
so it reduces traffic accidents. / It is also known / to reduce the severity of injuries /
所以减少了交通事故。/ 同时 / 减少受伤的严重程度 /
by a factor of three [1]. / Moreover, / street lighting reduces crime and fear of crime /
到原来的 1/3。/ 并且，/ 路灯减少犯罪和对犯罪的恐惧 /
and promotes the nighttime economy. /
以及促进夜间经济。/

Street lighting, / however, / has its own drawbacks. / It consumes a significant amount of electricity /
路灯，/ 但是，/ 有自身的不足。/ 它消耗大量电力 /
and requires continuous maintenance and repair, / all of which put pressure /
并且需要持续的维护和修理，/ 所有一切加重了负担 /
on a city's budget. / Street lighting is also a major source / of light pollution. /
给城市预算。/ 路灯也是主要的来源 / 光污染的。/
Light pollution, / or excessive artificial lighting, / is found / to have negative effects /
光污染，/ 或是过度的人工照明，/ 被发现 / 有不良影响 /
on both human and wildlife health [2].
对人类和野生动物的健康。/

That is why / cities / around the globe / are adopting Internet of Things technology /
这就是为什么 / 城市 / 全世界的 / 正采用物联网技术 /
for their streets. / Major cities have already started / to replace their existing street lights /
在它们的街道。/ 主要城市已经开始 / 替换现有路灯 /
with more efficient light-emitting diodes (LEDs). / New York plans / to replace 250,000 lights /
用更高效的 LED 灯。/ 纽约计划 / 替换 25 万个灯泡 /
by 2017 / and expects an annual savings of $14 million [3]. / Seoul will replace all public lighting /
到 2017 年。/ 期待每年缩减 1400 万美元费用。/ 首尔将替换所有公共照明 /
with LEDs / by 2018 / and expand the project / to the private sector / by 2030 [4]. /
用 LED 灯 / 到 2018 年 / 并且扩大计划 / 到民间部门 / 到 2030 年。/
In addition to energy-efficient LEDs, / cities are also connecting street lights and sensors /
除了高效的 LED 灯 / 城市也在连接路灯和传感器 /

through a network / to enable smart street lighting. / According to Telematics Wireless [5], /
通过网络 / 以让智能路灯使用。/ 据 Telematics Wireless 称，/
smart street lighting systems offer the following benefits: /
智能路灯系统有以下优点。/

They reduce maintenance costs. / The system detects and sends alerts for / lamp burnouts /
减少维护费用。/ 系统监视并且发出警告 / 对路灯故障 /
and other malfunctions / in real time, / which enables the maintenance crew /
和其他故障 / 实时地 / 方便维修人员 /
to take timely actions. / No need for routine maintenance checks. /
及时采取行动。/ 无需定期维护检查。/
They allow dimming-intensity control. / The system can automatically dim the street lights /
可调节亮度。/ 系统可自动降低路灯亮度 /
or turn them on/off / based on various conditions / set by operators / or triggered /
或者开关 / 基于各种不同条件 / 由操作人员设置 / 或者触发的 /
by sensors. / For example, / the lights are dimmed / when there's no traffic. /
通过传感器。/ 例如，/ 灯变暗 / 当没有车流时。/
This achieves major energy savings / as well as a reduction / in light pollution. /
这极大地减少了能源消耗 / 也减少了 / 光污染。/
They extend the lifespan / of street lights. / LEDs have a longer life /
延长了寿命 / 路灯的。/LED 寿命较长 /
than conventional street lamps. / Smart dimming extends their life / even longer, /
比传统路灯。/ 智能亮度调节使其寿命 / 更长，/
which reduces maintenance costs. /
这减少了维护费用。/
They offer real-time reporting and control. / The system collects real-time data /
它们提供实时报告和控制。/ 系统收集实时数据 /
such as voltage, current, power factor, and energy consumption /
例如电压、电流、功率因数和能源消耗 /
to help / operators manage the grid / in a smarter way.
以帮助 / 操作人员管理电网 / 以更加智能的方式。/
This is how / your city is becoming smarter. / The street lamp / you passed by / last night /
这就是方式 / 你的城市正变得越来越智能的。/ 路灯 / 你经过的 / 在昨晚 /
may be more intelligent / than you think! /
可能更加智能 / 比你想像的！/

物联网时代的一天（上）

A Morning in Your Life with the Internet of Things

In the world of the Internet of Things, technology and devices work for you. They make your day smarter. So what will be your typical morning with the Internet of Things?

6:00 AM

The alarm rings. It's earlier than usual because you have a 9:00 a.m. meeting with overseas clients. You didn't set the alarm. Communicating with your calendar, the smart alarm wakes you up on time [1].

As soon as you turn off the alarm, the coffee maker starts brewing. The aroma fills the house and makes you feel awake. Lights and temperatures are already adjusted to help you wake up.

You go out for a quick morning jog. While you run, your smartwatch records steps, distance, and calories. It also monitors your heart rate and breathing. When you come home, it automatically uploads the data to your fitness app.

8:00 a.m.

You leave the house for work. Sensing your departure, the house automatically enters power-saving mode. The heating adjusts, the lights switch off, and the appliances power down or switch off as necessary [2].

While driving to work, your car recommends the best route to avoid slowed traffic due to road work. You arrive at the office with enough time to get ready for the meeting. Of course, your smart office already knew your arrival time and set the temperature at your comfort level.

11:30 a.m.

You are stepping out of the office for a doctor's appointment. Waiting in the doctor's office, you watch the screen playing an ad for a jacket. The jacket looks familiar. You saw the exact model a few days ago while shopping at the mall. When you tried the jacket on, it sent a cookie to your smartwatch. Your smartwatch tells the smart TV in the waiting room, and the TV plays the jacket ad while you are waiting [3].

Your doctor is ready for you. She pulls out your daily electrocardiogram (ECG) data from your health record. Your fitness app already uploaded the data to the e-healthcare system. You have a family history of heart disease and had a small heart attack about a year ago. So your doctor wants to monitor you regularly. She is very pleased with your daily ECG data and advises you to continue with your current diet and exercise. You schedule the next appointment and leave the doctor's office feeling happy. After grabbing a quick lunch, you are back at the office.

Well, the day's not over yet. Your smart day continues!

出 处

1. SMART Alarm Clock, http://goo.gl/cysqJJ
2. WHAT WOULD YOU GIVE TO LIVE IN A SMART HOME? , http://goo.gl/S88K6s
3. 9 Real-Life Scenarios That Show How The Internet Of Things Could Transform Our Lives, http://goo.gl/AVDgBu

核心语法

学习#句式5〈主语+动词+宾语+宾补〉。作#宾补的动词形态根据与宾语的关系以及整个句子的动词而变化。例文中#宾语和#宾补为主动关系。make为#使役动词，后接#动词原形作#宾补。watch为#感官动词，后可接#动词原形和#现在分词作#宾补。

- They **make** your day **smarter**. 它让/你的一天更智能。
- The aroma **makes** you **feel** awake. 香气让/你感到/清醒。
- You **watch** the screen **playing** an ad. 你看到/屏幕在播放/广告。

熟悉#分词从句。#分词从句是缩短句子的方式之一，具体省略方式为：在#分词从句〈连词+主语+动词〉中去掉可推测的内容，或将动词变为分词。例文将介绍常用#分词从句。以下例句中，省略的内容并不影响原意。

- **Communicating** with your calendar, the smart alarm wakes you up on time. 和你的日历通信，智能闹钟叫醒你/按时。
  = **After the smart alarm communicates** with your calendar, the smart alarm wakes you up on time. =智能闹钟和你的日历通信后，智能闹钟叫醒你/按时。
- **Sensing** your departure, the house automatically enters power–saving mode. 感应到你出发，房子自动进入节电模式。
  = **As the house senses** your departure, the house automatically enters power–saving mode. =随着房子感应到你出发，房子自动进入节电模式。
- **While Driving** to work, your car recommends the best route. 驾车去工作，你的汽车推荐/最优路线。
  = **While your car drives** to work, your car recommends the best route. =当你驾车去工作，你的汽车推荐/最优路线。

- You leave the doctor's office **feeling** happy. 你离开医院/感到幸福。
  = You leave the doctor's office **and you feel** happy. =你离开医院/ 并且（同时）感到幸福。

熟悉连接两个句子的 # 连词。

- **As soon as** you turn off the alarm, ~ 一旦你关闭闹钟，
- **While** you run, ~ 在你跑步期间，

单词&短语

typical day 典型的一天
earlier than usual 早于平时
overseas client 国外客户
set an alarm 定闹钟
on time 按时
as soon as ~ 一 ~ 就
turn off 关闭
brew 冲泡（咖啡、茶）
aroma 香气
temperature 温度
adjust 调节
record 记录
step 步伐
distance 距离
heart rate 心率
breathing 呼吸
automatically 自动地
leave A for B 离开 A 前往 B
sense 感应
departure 出发
power-saving mode 节电模式
heating 加热
switch off 关闭
appliances 家用电器
power down 电源休眠
as necessary 必需
recommend 推荐
route 路线
due to ~ 因为 ~
arrive at ~ 到达 ~
get ready for ~ 准备 ~
arrival time 到达时间
comfort 舒适
step out of ~ 从 ~ 出来
appointment 约定
exact 准确的
try on 试穿
waiting room 候机室，候车室
pull out 拉出
electrocardiogram 心电图
family history 家族史
heart disease 心脏病

heart attack 心脏麻痹
regularly 定期地
be pleased with ~ 对 ~ 喜欢
advise A to V 建议 A 做 V
diet and exercise 饮食习惯和运动
schedule 制定日程
grab a quick lunch 快速吃午餐
be over 结束
continue 继续

技术术语

IPS（Indoor Positioning System）室内定位系统为 GPS 定位的辅助定位。常见的室内无线定位技术有 Wi-Fi、蓝牙、红外线、超宽带、RFID、ZigBee 和超声波。

根据提示完成句子

As soon as you ____ ___ ___ _____, / the _____ ____ ____ _______. /
一旦你关掉闹钟，/ 咖啡机开始冲咖啡。/

The ___ _____ _____ _______ / and makes / you ____ _____. /
香气溢满房间/ 并且让/ 你感到清醒。/

______ and ____________ are already ________ / to help /
灯光和温度已经调节/ 以帮助/

you ____ __. /
你醒来。/

You __ ___ / for a quick morning jog. / While you run, /
你走出去/ 进行快速晨跑。/ 在你跑步的时候，/

your smartwatch records _____, ________, and ________. /
你的智能手表记录步数、距离和热量。/

It also ________ your _____ ____ and ________. /
它也记录你的心率和呼吸。/

When you come home, / it automatically uploads the data /
当你回家时，/ 它自动上传数据/

to your fitness app. /
到你的健康应用。/

思考题

解答题

1 （理解）以下哪一项与物联网不相符？
ⓐ 家用厨房系统 ⓑ 互联汽车
ⓒ 可穿戴设备 ⓓ 模拟电话

2 （论述）如何确定人处在室内还是室外（准确地说是自己的家）？

讨论

1 例文中的哪些内容可以立即实现？

课堂总结

1 学习#句式5、#使役动词、#感官动词的概念和例句。
2 学习#分词从句的概念和#分词从句的构句方法。
3 学习常用#连词和#连接副词。

答案
1 **d**.（模拟电话很难通过数字技术实现通信。） 2 使用IPS（室内定位系统）准确掌握人的位置。

翻译

A Morning / in Your Life / with the Internet of Things

物联网时代的一天（上）

In the world / of the Internet of Things, / technology and devices work / for you. /
在世界 / 名为物联网的，/ 技术和设备运转 / 为你。/
They make / your day smarter. / So what will be your typical morning / with the Internet of Things?
它们让 / 你的生活更加智能 / 那么你的典型的一天将会如何呢？ / 和网络一起的

6:00 a.m. / 清晨6时

The alarm rings. / It's earlier than usual / because you have a 9:00 AM meeting /
闹钟响。/ 比平时略早 / 因为你上午 9 时有会议 /
with overseas clients. / You didn't set the alarm. / Communicating with your calendar, /
与国外客户。/ 你没有设置闹钟。/ 和你的日历通信后，/
the smart alarm wakes you up / on time [1].
智能闹钟叫醒你 / 按时。/

As soon as you turn off the alarm, / the coffee maker starts brewing. /
一旦你关闭闹钟，/ 咖啡机开始冲咖啡。/
The aroma fills the house / and makes / you feel awake. /
香气溢满房间 / 让 / 你感到清醒。/
Lights and temperatures are already adjusted / to help / you wake up. /
灯光和温度已经调节好 / 以帮助 / 你醒来。/

You go out / for a quick morning jog. / While you run, /
你走出去 / 迅速晨跑。/ 在你跑步的时候，/
your smartwatch records steps, distance, and calories. /
你的智能手表记录步数、距离和热量。/
It also monitors your heart rate and breathing. / When you come home, /
还记录你的心率和呼吸。/ 当你回家时，/
it automatically uploads the data / to your fitness app. /
它自动上传数据 / 到你的健康应用。/

8:00 a.m. / 上午8时

You leave the house / for work. /
你离开家 / 去工作。/
Sensing your departure, / the house automatically enters power-saving mode. /
监测到你离开，/ 房间自动进入节电模式。/
The heating adjusts, / the lights switch off, / and the appliances power down /
暖气开始调整 / 灯光关闭，/ 家用电器进入休眠状态 /

or switch off as necessary [2]. /
或者根据需要关闭。/
While driving to work, / your car recommends the best route / to avoid slowed traffic /
当你开车去工作时，/ 汽车给你推荐最优路线 / 以避免交通拥堵 /
due to road work. / You arrive at the office / with enough time / to get ready for the meeting. /
道路施工带来的。/ 你到达办公室 / 有足够的时间 / 准备会议。/
Of course, / your smart office already knew your arrival time / and set the temperature /
当然，/ 你的智能办公室已提早得知你的到达时间 / 并设置好温度 /
at your comfort level. /
按照让你感觉舒适的标准。/

11:30 a.m. / 上午11时30分

You are stepping out of the office / for a doctor's appointment. / Waiting in the doctor's office, /
你走出办公室 / 因为和医生有约。/ 在诊疗室等待的同时，/
you watch / the screen playing an ad for a jacket. / The jacket looks familiar. /
你看到 / 屏幕正在播放夹克广告。/ 这件夹克看起来很眼熟。/
You saw the exact model / a few days ago / while shopping / at the mall. /
你看过这款 / 几天前 / 购物的时候 / 在商场。/
When you tried the jacket on, / it sent a cookie / to your smartwatch. /
当你试穿夹克，/ 它发送 Cookie/ 到你的智能手表。/
Your smartwatch tells the smart TV / in the waiting room, / and the TV plays the jacket ad /
你的智能手表告诉智能 TV/ 在候诊室里，/ 并且 TV 播放夹克广告 /
while you are waiting [3]. /
在你等待时。/

Your doctor is ready for you. / She pulls out your daily electrocardiogram (ECG) data /
你的医生在等你。/ 她拿出每日 ECG（心电图）数据 /
from your health record. / Your fitness app already uploaded the data /
从你的健康记录中。/ 你的健康应用已上传数据 /
to the e-healthcare system. / You have a family history / of heart disease /
到电子健康管理系统。/ 你有家族史 / 心脏病的 /
and had a small heart attack / about a year ago. / So your doctor wants to monitor you regularly. /
和轻度心脏麻痹 / 大约 1 年前。/ 所以医生想定期观察。/
She is very pleased with your daily ECG data / and advises you /
她非常满意你的每日 ECG 数据 / 并且建议你 /
to continue with your current diet and exercise. / You schedule the next appointment /
继续保持现在的饮食习惯和运动。/ 你约了下次见面的日程 /
and leave the doctor's office / feeling happy. /
之后离开了诊疗室 / 感到很开心。/
After grabbing a quick lunch, / you are back at the office. /
迅速吃完午饭后，/ 你回到办公室。/
Well, / the day's not over yet. / Your smart day continues! /
好吧，/ 一天还没有结束。/ 你的智能的一天仍将继续！/

物联网时代的一天（下）

An Evening in Your Life with the Internet of Things

In the world of the Internet of Things, technology and devices work for you. They make your day smarter. So what will be your typical evening with the Internet of Things?

5:30 p.m.

Your workday is over, and you are ready to head home. Before leaving work, you select a recipe from your smart oven's recipe bank [1]. The oven sends the ingredient list to your smartphone, which checks the fridge to figure out which items you're out of. You stop by a local grocery and pick up everything you need.

Arriving in front of the house, you hop out of the car. Sensing the car, the garage door opens up automatically. The car drives itself in. As you approach the front door, the smart lock receives a signal from your smartphone and unlocks the door [2].

The house is cool and comfortable. Lights and temperature are already set to your comfort level. The music you were listening to in the car continues playing in the living room.

7:00 p.m.

You start making dinner. Following the recipe you selected, you put the dish into the oven. There's no need to set the temperature. The oven adjusts it as necessary. It says dinner will be ready in 30 minutes [1].

While waiting, you decide to walk your dog. As soon as you walk out the door, your dog jumps and makes you drop its leash. The dog runs away, but you find it easily since you have a GPS pet tracker attached to the dog's collar [3].

When you arrive back home, your dinner is ready.

10:00 p.m.

It's time for bed. Knowing your audio habits, the smart music system selects and plays white noise. Today it's the sound of rain. While you sleep, your smart bed with tiny built-in sensors collects data on your sleeping habits. The bed automatically adjusts air pockets, helping you have a perfect sleep [2].

Since you didn't get enough sleep the night before, your bed tells your alarm to give you an extra hour of sleep. Your alarm checks your calendar to make sure you have no early meetings. You don't, so it will let you sleep another hour in the morning [4].

Now your smart day's over. What do you think? Are you excited about the coming Internet of Things revolution?

出 处

1. Smart Oven, http://goo.gl/NM9rth
2. A DAY WITH THE INTERNET OF THINGS, http://goo.gl/b9Rw51
3. The Internet of Things: Connected animals, http://goo.gl/GLx7vj
4. 9 Real-Life Scenarios That Show How The Internet Of Things Could Transform Our Lives, http://goo.gl/AVDgBu

核心语法

熟悉 # 句式 5〈主语 + 动词 + 宾语 + 宾补〉。# 动词原形、# 不定式 to、# 现在分词作 # 宾补时，与 # 宾语为主动关系；# 过去分词作 # 宾补时，与 # 宾语为被动关系。以下例句中的 have 为 # 使役动词，意为“让”“使得”。# 宾补表被动关系时，意为“使 ~ 处于某种状态”。

- Your bed tells your alarm to give you an extra hour of sleep. 你的床告诉/闹钟让你多睡一小时。
- It makes you drop its leash. 这让/你丢了它的绳子。
- You have a GPS pet tracker attached to the dog's collar. 你把/GPS宠物追踪器拴在狗脖子上。（你让GPS宠物追踪器被拴在狗脖子上。）

熟悉 # 分词从句。# 分词从句是缩短句子的方式之一，具体省略方式为：在 # 分词从句〈连词 + 主语 + 动词〉中去掉可推测的内容，或者将动词变为分词。例文将介绍常用 # 分词从句。以下例句中，省略的内容并不影响原意。

- **Leaving** work, you select a recipe. 下班前，你选择了一份食谱。
  = **Before you leave** work, you select a recipe. =在你下班前，你选择了一份食谱。
- **Sensing** the car, the garage door opens up automatically. 感应到车后，车库门自动打开。
  = **As soon as the garage door sense** the car, the garage door opens up automatically. =一旦车库感应到车，车库门自动打开。
- The bed adjusts air pockets, **helping** you have a perfect sleep. 床调整气袋， 帮你改善睡眠质量。
  = The bed automatically adjusts air pockets, **and the bed helps** you have a perfect sleep. =床自动调整气袋，并且（同时）帮你改善睡眠质量。

熟悉 # 关系代名词的省略。例文中省略 # 宾格关系代名词。省略后不影响原意。

- You buy **everything** ([**that**] you need). 你购买所有东西/你需要的。
- **The music** ([**that**] you were listening to in the car) continues playing in the living room. 音乐/你在车里听过的/继续在客厅播放。

熟悉 # 关系代名词引导非限定性定语从句。用于补充说明 # 先行词。# 关系代名词前加逗号（，）可理解为〈连词 + 代名词〉。

- The oven sends the list to **your smartphone**, **which** checks the fridge.
  = The oven sends the list to **your smartphone**, **and it** checks the fridge. 烤箱传输列表/到你的智能手机，/它（=你的智能手机）检查冰箱。

单词&短语

typical 典型的
over 结束
ready to V 准备做 ~
head ~ 前往 ~
recipe 食谱
ingredient 材料，成分
fridge 冰箱
figure out 想出，弄明白
be out of 短缺
stop by 顺便走访
local grocery 当地杂货店
arrive 到达
hop out of ~ 跳出 ~
sense 感应
garage door 车库门
automatically 自动地
approach 接近
receive 接收
unlock 打开，解锁
comfortable 舒适的
temperature 温度
continue 继续
select 选择
adjust 调节
walk a dog 遛狗
leash 狗绳
pet tracker 宠物追踪器
attach 贴，附

collar 项圈

built-in sensors 内置传感器

be excited about ~ 对 ~ 感到兴奋

revolution 革命

技术术语

white noise 白噪声是一种功率频谱密度为常数的随机信号或随机过程。换言之，此信号在各频段上的功率是一样的。由于白光是各种频率（颜色）的单色光混合而成的，所以此信号的这种具有平坦功率谱的性质被称为“白色的”，故而得名。

根据提示完成句子

Arriving __ ______ __ the house, / you __ ___ __ the car. /
到家门前，/ 你下车。/

_______ the car, / the ______ ____ _____ __ / automatically. /
感应到车，/ 车库门自动打开。/

The car _______ itself __. / As you _________ ___ _____ ____, /
车自动驶入。/ 当你接近前门，/

___ _____ ____ ________ _ ______ /
智能锁接收到信号/

from your smartphone / and _______ the door [2]. /
从你的智能手机/ 然后解锁。/

The house is ____ and __________. /
房间凉爽舒适。/

______ and ____________ are already set / to your _______ _____. /
灯光和温度已经设定/ 符合让你舒适的标准。

The music you were listening to / in the car / __________ _______ /
你听的音乐/ 在车里/继续播放/

in the _______ ____. /

在客厅。/

思考题

解答题

1 （理解）以下哪一项与物联网不相符？

ⓐ 智能微波炉　　ⓑ 停车场大门

ⓒ 狗链　　ⓓ 床

2 （论述）如何确定冰箱内有哪种食材？请指出此种方法的缺陷。

讨论

1 例文中的哪些内容可以立即实现？

课堂总结

1 学习#句式5、#使役动词、#感官动词的概念和例句。

2 学习#分词从句的概念和#分词从句的构句方法。

3 学习常用#连词和#连接副词。

答案

1 c. 2 将RFID标签粘贴在冰箱上就可掌握冰箱内的食材类型（缺点是不能准确把握量）。通过计算机可视技术使用相机可以掌握食材类型（缺点是不能准确区分）。

翻译

An Evening / in Your Life / with the Internet of Things
物联网时代的一天（下）

In the world / of the Internet of Things, / technology and devices work / for you.
在世界 / 名为物联网的，/ 技术和设备运转 / 为你。/
They make / your day smarter. / So what will be your typical evening / with the Internet of Things?
它们让 / 你的生活更加智能 / 那么你的典型的一天将会如何呢？ / 和网络一起的

5:30 p.m. / 下午5时30分

Your workday is over, / and you are ready to head home. / Before leaving work, /
你结束了工作，/ 准备回家。/ 下班之前，/
you select a recipe /
你选择了一份食谱 /
from your smart oven's recipe bank [1]. The oven sends the ingredient list / to your smartphone, /
从你的智能微波炉食谱库中。/ 微波炉传输食材列表 / 到你的智能手机，/
which checks the fridge / to figure out / which items you're out of. / You stop by a local grocery /
它检查冰箱 / 确认 / 哪些食材短缺。/ 你去当地杂货店 /
and pick up everything / you need. /
选出所有 / 你需要的东西。/

Arriving in front of the house, / you hop out of the car. / Sensing the car, / the garage door opens up /
到家门前，/ 你下车。/ 感应到车 / 车库门打开 /
automatically. / The car drives itself in. / As you approach the front door, /
自动地。/ 车自动驶入。/ 当你接近前门，/
the smart lock receives a signal / from your smartphone / and unlocks the door [2]. /
智能锁接收信号 / 从你的智能手机 / 然后解锁。/

The house is cool and comfortable. / Lights and temperature are already set / to your comfort level. /
房间凉爽且舒适。/ 灯光和温度已经设定 / 符合让你舒适的标准。/
The music you were listening to / in the car / continues playing / in the living room. /
你听的音乐 / 在车里 / 继续播放 / 在客厅。/

7:00 p.m. / 晚上7时

You start making dinner. / Following the recipe you selected, / you put the dish into the oven. /
你开始做晚饭。/ 按照你选择的食谱，/ 你把盘子放进微波炉。/

There's no need / to set the temperature. / The oven adjusts it / as necessary. /
没必要 / 设定温度。/ 微波炉调整温度 / 根据需要。/
It says / dinner will be ready in 30 minutes [1]. /
它说 / 晚餐会在 30 分钟内准备好。/

While waiting, / you decide / to walk your dog. / As soon as you walk out the door, / your dog jumps /
在等待的时间里，/ 你决定 / 遛狗。/ 一出门，/ 狗开始跳 /
and makes you drop its leash. / The dog runs away, / but you find it easily /
使你丢了狗链。/ 狗跑掉了，/ 但你很容易找到它 /
since you have a GPS pet tracker / attached to the dog's collar [3]. /
因为你系上了 /GPS 宠物追踪器在狗脖子上。/

When you arrive back home, / your dinner is ready. /
当你回家时，/ 你的晚餐已经准备好。/

10:00 p.m. / 晚上10时

It's time for bed. / Knowing your audio habits, / the smart music system selects /
到了睡觉时间。/ 了解到你听音乐的习惯，/ 智能音乐系统开始选择 /
and plays white noise. /
并且播放白噪声。/
Today / it's the sound of rain. / While you sleep, / your smart bed / with tiny built-in sensors /
今天 / 是雨声。/ 在你睡觉的时间里，/ 你的智能床 / 带有小型内置传感器的 /
collects data / on your sleeping habits. / The bed automatically adjusts air pockets, /
搜集数据 / 关于你的睡眠习惯。/ 床自动调节气袋，/
helping you have a perfect sleep [2]. /
以帮助你改善睡眠。/

Since you didn't get enough sleep / the night before, /
因为你没睡好 / 昨天晚上，/
your bed / tells your alarm to give you an extra hour of sleep. /
你的床 / 通知闹钟让你多睡 1 小时。/

Your alarm checks your calendar / to make sure / you have no early meetings. / You don't, /
闹钟检查你的日历 / 以确认 / 你没有早会。/ 没有，/
so it will let you sleep another hour / in the morning [4]. /
所以它会让你多睡 1 小时 / 早上。/
Now your smart day's over. / What do you think? / Are you excited /
现在你的智能的一天结束了。/ 你认为如何？兴奋吗？ /
about the coming Internet of Things revolution? /
对于即将到来的物联网革命？ /

互联汽车

A Connected Car Saves You Time and Money

A connected car is a car with Internet access as well as a wireless local area network [1]. This allows the car to share Internet access with other devices both inside and outside the vehicle. Often, connected cars come with music/audio playing, navigation, roadside assistance, voice commands, parking aid, engine controls, and problem diagnosis. The newer models take advantage of smartphone apps to allow interaction with the car from anywhere. Drivers can remotely unlock their cars, find the location of their cars, or activate their climate control systems. But connected cars bring drivers more than just convenience. They help save time, money, the environment, and even lives [2].

More Efficient Traffic

Connected cars can utilize real-time traffic information to help traffic move more efficiently. San Antonio, Texas, uses connected car technology and GPS to keep the city's bus system on schedule and improve traffic flow. According to a U. S. Department of Transportation (DOT) study, intersections with connected vehicle technology can lower traffic delays by 17 percent.

Less Fuel Consumption

Connected cars use less gasoline. Better traffic flow means fewer stops at intersections. Fewer stops on roads lead to less gasoline consumption. Less gasoline usage translates to less CO_2 emissions.

"According to recent statistics from Nationwide Insurance, the average urban commuter gets stuck in traffic an estimated 34 hours every year, and we waste 1.9 billion gallons of fuel," said Ben Collar, director of research and development for Siemens Road and City Mobility in Austin, Texas.

Fewer Accidents

Connected cars can help avoid accidents by directly communicating with other cars. According to the same DOT study, the use of connected vehicle technology can help reduce car crashes by 83%.

"Soon cars will be able to communicate with each other to avoid collisions," Collar said. "The same will work for vehicle-to-infrastructure technology as cars will be able to communicate with roads and detect hazards such as pedestrians and downed trees."

Money Savings

All these factors help drivers spend less time on roads and less money on fuel. Fewer accidents will result in lower auto insurance rates. Connected software can offer drivers additional savings.

"Manufacturers and service providers can remotely manage the embedded software in cars throughout the lifetime of the vehicle," said Roger Ordman from the software company, Red Bends. "They can also substantially lower software-related warranty costs and avoid the expensive direct and indirect costs associated with product recalls."

出 处

1. Connected car, Wikipedia, http://goo.gl/vTI3zQ
2. How a "connected" car could save you money, http://goo.gl/NzGkTK

核心语法

学习 # 句式 3〈主语 +help+ 宾语〉。help 后可以接 # 动词原形或者 # 不定式 to 作 # 宾语。

- They help [to] **save** time. 它们有助于/节约时间。
- They can help [to] **avoid** accidents. 它们有助于/避免交通事故。
- The use of technology can help [to] **reduce** car crashes by 83%. 使用技术有助于/减少车辆碰撞/83%。

熟悉 # 句式 4〈主语 + 动词 + 间接宾语 + 直接宾语〉，# 句式 4 中使用的动词为 # 授予动词（双宾动词），意为“使~，让~”。

- A connected car **saves** you time. 互联汽车节约/让你/时间。
- Connected cars **bring** drivers convenience. 互联汽车带来/给司机/方便。
- Connected software can **offer** drivers additional savings. 互联汽车能提供/给司机/更多节约。

学习 # 句式 5〈主语 + 动词 + 宾语 + 宾补〉。例文将介绍〈主语 + 谓语〉关系的 # 宾语和 # 宾补。此时 # 宾补的形态根据 # 动词的变化而变化。allow 后加 # 不定式 to 作 # 宾补。help 后可接 # 动词原形和 # 不定式 to。在阅读和听力训练中，只要掌握 # 宾语和 # 宾补的主谓关系即可；但在写作和会话训练中，还必须掌握各动词后接的 # 宾补的形态。

- This **allows** the car **to share** Internet access with other devices. 这让/车共享/网络/与其他设备。
- Connected cars **help** traffic **move** more efficiently. 互联汽车有助于/让车流移动/更加顺畅。
- These factors **help** drivers **spend** less time on roads. 这些要素有助于/司机花费/更少时间/在路上。

熟悉与数值相关的 # 前置词 by 的用法。by 表示手段或者方法时，意为“用～”；表示数值增加时，意为“～程度”。

- Connected cars can help avoid accidents **by** directly communicating with other cars. 互联汽车有助于/避免交通事故/通过与其他车辆直接通信的方法。
- It can lower traffic delays **by** 17 percent. 这能降低/交通堵塞/达17%。
- The use of connected vehicle technology can help reduce car crashes **by** 83%. 互联汽车技术的使用有助于/减少车辆碰撞/达83%。

单词&短语

B as well as A 不仅 A 而且 B
allow ~ to V 允许 ~ 做 V
share A with B 和 B 共享 A
come with ~ 伴随 ~
navigation 导航
roadside assistance 道路救援
parking aid 停车辅助
diagnosis 诊断
take advantage of ~ 利用 ~
interaction with ~ 和 ~ 沟通
remotely 远程
unlock 解锁
activate 启动
climate control system 汽车空调系统
convenience 方便，便利
environment 环境
efficient 有效的
utilize 使用
real-time traffic information 实时交通信息
on schedule 按计划
traffic flow 车流量
U.S. department of transportation 美国交通部
intersection 十字路口
traffic delay 交通堵塞
fuel consumption 燃料消耗
gasoline usage 耗油量
lead to ~ 导致 ~
translate to ~ 将 ~ 转换为
CO_2 emission 二氧化碳排放量
statistics 统计，统计学
average urban commuter 普通上班族
get stuck 卡住
estimated 估计
waste 浪费
research and development 研发

技术术语

M2M（Machine to Machine）将数据从一台终端传送到另一台终端，也就是机器与机器的对话。互联汽车不仅需要车与车之间通信，还需要车与基础设施之间通信，主要使用 M2M 通信技术。

根据提示完成句子

A connected car is a car / with ________ ______ /
互联汽车是一种车/ 带有网络/

as well as a ________ _____ ____ _______. /
及无线局域网。/

This allows the car / to _____ _________ _______ / with other devices /
这让车辆/ 可以共享网络/ 与其他设备/

____ ______ ___ _______ ___ _______. /
在车辆内外部。/

Often, / connected cars ____ ____ / music audio playing, / navigation, /
通常，/ 互联汽车具有/ 音乐播放、/ 导航、/

_________ ___________, / voice commands, / _______ ___, /
道路救援、/ 语音控制、/ 停车辅助、/

engine controls, / and _______ _________. /
引擎控制/和车辆诊断。

The newer models take advantage /
较新车型利用/

of smartphone apps / to allow ___________ ____ the car /
智能手机应用/ 与车辆通信/

____ ________. / Drivers can ________ ______ their cars, /
在任何地方。/ 司机可以远程解锁车辆，/

____ ___ _________ of their cars, /
找到车辆的位置，/

or ________ _____ _______ ______ ________. /
或者控制空调系统。/

But connected cars bring drivers / ____ ____ ____ ____________. /
但互联汽车带给司机的/ 不仅仅是方便。/

They help / save / time, money, the environment, and even lives [2]. /
它们有助于/节约/时间、金钱，保护环境甚至生命。/

思考题

解答题

1 （理解）以下哪一项不是互联汽车的优点？

ⓐ 高效的车辆移动　ⓑ 节约能源

ⓒ 预防事故　ⓓ 车价上涨

2 （理解）以下哪一项不是互联汽车的功能？

ⓐ 导航　ⓑ 停车辅助

ⓒ 车辆诊断　ⓓ 赶走小偷

3 （论述）简述互联汽车的经济效益。

4 （论述）简述互联汽车减少交通事故的方法。

讨论

1 最近特斯拉宣布，只需升级软件，其旗下的特斯拉Model S即可增加高速公路自动驾驶功能。请简述这种与iPhone升级相关车型的优缺点。

2 请思考在车辆共享服务中，互联汽车受追捧的原因。

课堂总结

1 学习#不定式和#动名词作#宾语的#动词和例句。

2 学习#句式4和#授予动词例句。学习#句式3的转换。

3 学习#句式5中#宾补的形态和相关动词。

答案

1 d. 2 d. 3 节约燃料以节省费用。交通顺畅后节约时间（费用），事故减少后节约保险费（以及社会成本），允许软件下载，减少了产品召回成本。 4 因车辆之间可直接通信，所以可随时应对交通事故的发生。并且车辆和周边基础设施可以通信，能提前应对行人和障碍物等。

翻译

A Connected Car Saves You Time and Money
互联汽车

A connected car is a car / with Internet access / as well as a wireless local area network [1]. /
互联汽车是一种车 / 带有网络 / 及无线局域网。/
This allows the car / to share Internet access / with other devices / both inside and outside the vehicle. /
这让车辆 / 可以共享网络 / 与其他设备 / 在车辆内外部。/
Often, / connected cars come with / music audio playing, / navigation, /
通常，/ 互联汽车具有 / 音乐播放、/ 导航、/
roadside assistance, / voice commands, / parking aid, / engine controls, /
道路救援、/ 语音控制、/ 停车辅助、/ 引擎控制 /
and problem diagnosis. / The newer models take advantage / of smartphone apps /
以及车辆诊断。/ 较新车型利用 / 智能手机应用 /
to allow interaction with the car / from anywhere. / Drivers can remotely unlock their cars, /
与车辆通信 / 在任何地方。/ 司机可以远程解锁车辆，/
find the location of their cars, / or activate their climate control systems. /
找到车辆的位置 / 或者控制空调系统。/
But connected cars bring drivers / more than just convenience. / They help / save /
但互联汽车给司机带来的 / 不仅仅是方便。/ 它们有助于 / 节约 /
time, money, the environment, and even lives [2]. /
时间、金钱，保护环境甚至生命。/

More Efficient Traffic / 更顺畅的交通

Connected cars can utilize real-time traffic information / to help traffic move /
互联汽车利用实时交通信息 / 让车辆移动 / 更加顺畅。/
more efficiently. / San Antonio, / Texas, / uses connected car technology /
圣安东尼奥 / 得克萨斯，/ 使用互联汽车技术 /
and GPS to keep the city's bus system / on schedule / and improve traffic flow. /
和 GPS/ 维护城市的公交车系统 / 预计 / 并且改善交通状况。/
According to a U.S. Department of Transportation (DOT) study, /
依据美国交通部调查，/
intersections with connected vehicle technology / can lower traffic delays / by 17 percent. /
使用互联汽车技术的十字路口 / 能降低交通拥堵 / 达 17%。/

Less Fuel Consumption / 较少的燃料消耗

Connected cars use less gasoline. / Better traffic flow means fewer stops /
互联汽车使用较少的汽油。/ 更顺畅的交通意味着可以少停车 /
at intersections. / Fewer stops / on roads / lead to less gasoline consumption. /
在十字路口。/ 较少停车 / 在路上 / 导致汽油消耗较少。/

Less gasoline usage translates / to less CO_2 emissions. /
汽油消耗变少 / 意味着二氧化碳排放量变少。/
“According to recent statistics / from Nationwide Insurance, /
“据最新统计 / 来自全美互惠保险的，/
the average urban commuter gets stuck / in traffic / an estimated 34 hours / every year, /
普通上班族被堵 / 在路上 / 约 34 小时 / 每年，/
and we waste 1.9 billion gallons of fuel,” / said Ben Collar, /
并且我们浪费了 19 亿加仑的汽油，” / 本・科勒说，/
director of research and development for Siemens Road and City Mobility / in Austin, / Texas. /
西门子道路与城市交通研发主任 / 位于奥斯丁的 / 德克萨斯。

Fewer Accidents / 更少的交通事故

Connected cars can help / avoid accidents / by directly communicating / with other cars. /
互联汽车有助于 / 避免交通事故 / 通过直接通信 / 和其他车辆。/
According to the same DOT study, / the use of connected vehicle technology can help /
根据美国交通部的同一项调查，/ 互联汽车技术的使用有助于 /
reduce car crashes / by 83%. /
减少车辆碰撞 / 达 83%。/
“Soon cars will be able to communicate / with each other / to avoid collisions,” /
“很快车辆将能通信 / 彼此之间 / 以避免碰撞，” /
Collar said. / “The same will work / for vehicle-to-infrastructure technology /
科勒说。/ “同样的技术将会适用于 / 车辆和基础设施 /
as cars will be able to communicate / with roads / and detect hazards /
因为车辆将能够通信 / 与道路 / 并感知危险 /
such as pedestrians / and downed trees.” /
例如行人 / 和倒下的树木。” /

Money Savings / 节约费用

All these factors help drivers / spend less time / on roads / and less money / on fuel. /
所有因素都有助于司机 / 节省时间 / 在路上的 / 和费用消耗 / 在燃料上的。/
Fewer accidents will result in lower auto insurance rates. /
较少的交通事故将导致较少的保险费用。/
Connected software can offer drivers / additional savings. /
互联汽车能提供 / 给司机 / 更多节约。/

“Manufacturers and service providers / can remotely manage the embedded software /
“制造商和服务供应商 / 可远程管理嵌入式软件 /
in cars / throughout the lifetime / of the vehicle,” / said Roger Ordman /
车内的 / 在整个使用寿命期内 / 车辆的，” / 罗杰・奥德曼说 /
from the software company, / Red Bends. /
来自软件公司，/Red Bends。/
“They can also substantially lower software-related warranty costs /
“它们也可以相当程度上降低软件相关的保证金 /
and avoid the expensive direct and indirect costs associated / with product recalls.”
并且避免直接和间接的高额费用 / 与产品召回相关的。” /

衬衫预警心脏麻痹

Shirt Warns You of Heart Attack

A heart attack is one of the most common causes of sudden death. Even if it doesn't cause death, a heart attack can result in permanent heart damage and life-threatening problems. Every year, more than 19 million people worldwide die from a heart attack. In the United States alone, about 1.4 million people experience a heart attack each year [1].

Most heart attacks can be treated if you get medical assistance right away. However, people often don't get help in time and miss the window of opportunity. Why? It's because people often fail to recognize the early warning signs. It's because symptoms may be mild and easy to ignore.

But what if your shirt can detect these symptoms? What if it can warn you to go to hospital immediately? What if it can transmit your electrocardiogram (ECG) data to your smartphone or maybe even to your doctor?

At the IoT Connect 14 conference in Sydney, during his speech there, Dr. Hugh Bradlow talked about the potential of the Internet of Things (IoT) to transform healthcare and save millions of lives. He said that Internet-connected wearable devices can help people react faster to a medical emergency [2].

Companies around the globe are already moving toward such wearables [3]. For example, the Canadian tech firm OMsignal offers a wide range of smart fitness clothing and gadgets [4]. The company's wearables have built-in sensors to track the wearer's important biometric data and vital signs such as heart rate, breathing, movement, and calories. The sensors relay this real-time information to a smartphone app. These products can monitor workouts and detect stress. They are designed to help the wearer manage his or her health better. Future versions will be able to detect potentially dangerous anomalies before emergencies occur.

This utopian world has its hurdles, however. Dr. Bradlow points out that anomaly detection is very tricky. One may feel some symptoms early in the morning and have a heart attack much later in the afternoon. Furthermore, Bluetooth instability, smartphone viruses, bugs and other faults, and human error are also potential risks.

Still, the technology is advancing fast. The future may be closer than you think. Soon, if you are stressed too much or your heartbeat is irregular, your shirt may tell you to take rest or to seek medical help immediately. Just be sure to listen to your shirt's advice!

出 处

1. What You Should Know, Safety for Heart Attack Prevention and Eradication, http://goo.gl/OKloY4
2. Telstra looks to health benefits of connected devices, http://goo.gl/pWvWmX
3. Gussied up with smart fashion, http://goo.gl/QVgjtj
4. OMsignal Biometric Smartwear, http://www.omsignal.com

核心语法

熟悉 # 句式 5〈主语 + 动词 + 宾语 + 宾补〉。作 # 宾补的动词形态根据与宾语的关系以及整个句子的动词而变化。例文介绍的 # 宾语和 # 宾补为〈主语 + 谓语〉关系。warn 和 tell 的 # 宾补是 # 不定式 to。help 后可接 # 动词原形和 # 不定式 to。

- It can **warn** you **to go** to hospital. 它会提醒你/去医院。
- The devices can **help** people [**to**] **react** faster to a medical emergency. 设备有助于/人们更快应对/急诊。
- Your shirt may **tell** you **to take** rest. 你的衬衫可能会提示你/休息。

熟悉 # 不定式 to 用作 # 名词、# 形容词、# 副词的句子。根据上下文理解意思即可。注意，因为以上 # 句式 5 中都是将 # 不定式用作 # 宾补，所以皆为 # 名词性用法。

- People often fail **to recognize** the early warning signs. 人们经常不能/察觉早期征兆。
- He talked about the potential **to transform** healthcare. 他谈论潜力/将改变医疗服务的。
- Symptoms are easy **to ignore**. 症状容易/被忽视。
- They have built-in sensors **to track** the wearer's biometric data. 它们用内置传感器/追踪用户的身体信号。
- 它们用内置传感器/为了追踪用户的身体信号。

熟悉 # 并列句。# 并列句由两个或两个以上并列而又独立的简单句或句子成分构成，最后一句或一项前加 # 并列连词 and、or、but。# 并列句中也可以套用 # 并列句。

- Sensors track vital signs such as **heart rate**, **breathing**, **movement**, **and**

calories. 传感器追踪身体信号/例如心率、呼吸、移动、热量。

- **Bluetooth instability**, **smartphone viruses**, **bugs**, **and human error** are also potential risks. 不稳定的蓝牙、智能手机病毒、Bug、人类过失都是潜在的风险。
- Bluetooth instability, smartphone viruses, **bugs and other faults**, and human error are also potential risks. 不稳定的蓝牙、智能手机病毒、Bug和其他问题，以及人类过失都是潜在的风险。

单词&短语

warn A of B 警告 A 注意 B
heart attack 心脏麻痹
common cause 常见原因
sudden death 猝死
result in ~ 导致 ~
permanent damage 永久损伤
life-threatening 威胁生命
die from ~ 死于 ~
Experience 经历
treat 治疗
medical assistance 医疗救助
in time 及时
miss an opportunity 错失机会
window of opportunity 绝好的机会
fail to ~ 没能做 ~，失败做 ~
recognize 识别
early warning sign 早期预警信号
symptoms 症状
mild 轻微的，温柔的
ignore 忽视
detect 监测
warn A to V 警告 A 做 V
what if ~ 如果 ~
transmit 传送
electrocardiogram 心电图
conference 会议
speech 演讲
potential 潜力
transform 转换
save lives 拯救生命
millions of 数百万的
wearable device 可穿戴设备
react to ~ 对 ~ 做出反应
medical emergency 急诊
around the globe 全球的
move towards ~ 朝向 ~ 移动
tech firm 技术公司
offer 提供
a range of 各种各样的
clothing 服装

技术术语

IoT（Internet of Things）物联网通过射频识别（RFID）、红外感应器、全球定位系统、激光扫描仪等信息传感设备，按约定的协议，将一切物品通过物联网域名相连接，进行信息交换和通信，以实现智能化识别、定位、跟踪、监控和管理。通过物联网收集到的信息量堪比大数据。

根据提示完成句子

But / ____ __ / ____ _____ can ______ _____ ________? /

但是/ 会怎样？/ 你的衬衫可以监测这种症状/

____ __ / it can ____ ___ /

会怎样？/ 它可以提醒你/

to go to hospital / immediately? / ____ __ /

去医院/ 立刻？/ 会怎样？/

it can _________ ____ ___________ (ECG) data /

它能传送你的心电图（ECG）数据/

to your smartphone / or maybe even to your doctor? /

到你的智能手机/或者给你的医生？/

Still, / the technology is ___________ ____. /

仍然，/ 技术在飞速进步。/

The future may be closer / than you think. /

未来可能离我们更近/比你想的。/

Soon, / if you ___ _________ / too much /

很快，/ 如果你压力/ 过多/

or your _________ is __________, / your shirt may ____ you /

或者心率不齐，/ 你的衬衫可能会告诉你/

to ____ ____ / or to ____ _______ ____ / immediately. /

休息/ 或者寻求医疗救助/ 立即。/

Just __ ____ / to ______ __ ____ ______ ______! /

切记/ 要遵从衬衫的建议！/

思考题

解答题

1 （理解）以下哪一项最可能是人们因心脏麻痹死亡的原因？

ⓐ 心脏麻痹没有预警，属突发事件

ⓑ 人们对自己的健康过度自信

ⓒ 不关注健康

ⓓ 心脏麻痹的早期症状很难发现

2 （理解）以下哪一项为心脏麻痹存在危险性的原因？

ⓐ 会导致突发死亡　　ⓑ 一旦患病完全无法治愈

ⓒ 早期症状完全不显现　　ⓓ 非常少见的疾病

3 （论述）请列举可收集到的关于预防心脏麻痹的身体数据。

4 （论述）请指出可穿戴设备如何将收集到的数据传达给主治医生或者医院。

讨论

1 预防心脏麻痹的技术本身存在缺陷时可能会有副作用，请列举解决方法。

2 使用云模型可汇总分析多人的心脏麻痹身体数据。请简述在此过程中需要考虑的社会/经济因素。

课堂总结

1 学习#句式5、#使役动词、#感官动词的概念和例句。

2 学习#不定式to作#名词、#形容词、#副词的例句。

3 学习#并列句、#并列连词、#关联连词的概念和例句。

答案

1 d. 2 a. 3 心率、呼吸、移动、热量消耗 4 利用内置的调制解调器直接传输或者使用连接蓝牙的手机应用传输

翻译

Shirt Warns You / of Heart Attack
衬衫预警心脏麻痹

A heart attack is one of the most common causes / of sudden death. /
心脏麻痹是常见原因之一 / 猝死的。/
Even if it doesn't cause death, / a heart attack can result in permanent heart damage /
即使不会导致死亡 / 心脏麻痹也能引发永久性心脏损伤 /
and life-threatening problems. / Every year, / more than 19 million people / worldwide /
并威胁生命。/ 每年，/ 超过 1900 万人 / 全球 /
die from a heart attack. / In the United States alone,/ about 1.4 million people /
死于心脏麻痹。/ 仅在美国，/ 就有 140 万人 /
experience a heart attack / each year [1]. /
患心脏麻痹 / 每年。/

Most heart attacks can be treated / if you get medical assistance / right away. /
多数心脏麻痹可以治愈 / 如果就医 / 及时。/
However, / people often don't get help / in time / and miss the window of opportunity. /
但是，/ 人们通常就医 / 不及时 / 并且错过了最佳时机。/
Why? / It's because / people often fail / to recognize the early warning signs. /
原因何在？ / 因为 / 人们通常无法 / 察觉早期症状。/
It's because / symptoms may be mild / and easy / to ignore. /
因为 / 症状不明显 / 并且容易 / 被忽视。/

But / what if / your shirt can detect these symptoms? / What if / it can warn you /
但是 / 会怎样呢？ / 你的衬衫可以监测这种症状？ / 会怎样呢？ / 它可以提醒你 /
to go to hospital / immediately? / What if / it can transmit your electrocardiogram (ECG) data /
去医院 / 及时地？ / 会怎样呢？ / 它能传送你的心电图（ECG）数据 /
to your smartphone / or maybe even to your doctor? /
到你的智能手机 / 或者你的医生？ /

At the IoT Connect 14 conference / in Sydney, / during his speech there, / Dr. Hugh Bradlow talked /
在 IoT Connect 14 会议上 / 在悉尼召开，/ 在他的演讲中，/ 休·布莱德劳博士谈了 /
about the potential of the Internet of Things (IoT) / to transform healthcare /
关于 IoT 的潜力 / 在改变医疗服务 /
and save millions of lives. / He said / that Internet-connected wearable devices can help people /
和拯救数百万人的生命。/ 他说 / 联网的可穿戴设备能帮助人们 /
react faster / to a medical emergency [2]. /
快速应对 / 急诊。/

Companies / around the globe / are already moving toward such wearables [3]. /
公司 / 全球的 / 已经开始向可穿戴设备领域采取行动。/
For example, / the Canadian tech firm OMsignal /
例如，/ 加拿大技术公司 OMsignal/
offers a wide range of smart fitness clothing and gadgets [4]. /
提供众多智能运动服和小配置。/
The company's wearables have built-in sensors / to track the wearer's important biometric data /
这家公司的可穿戴设备内置了传感器 / 以追踪用户重要的身体数据 /
and vital signs / such as heart rate, breathing, movement, and calories. /
和身体信号 / 例如心率、呼吸、运动、热量等。/
The sensors relay this real-time information / to a smartphone app. /
传感器实时传输信息 / 到智能手机应用。/
These products can monitor workouts / and detect stress. / They are designed /
这些产品能监测产品 / 感知压力。/ 它们的设计 /
to help the wearer / manage his or her health better. /
目的是帮助用户 / 更好地管理自己的健康。/
Future versions will be able to detect potentially dangerous anomalies /
今后的版本将能感知潜在的危险异常 /
before emergencies occur. /
在紧急状况发生之前。/

This utopian world has its hurdles, / however. / Dr. Bradlow points out /
这种“乌托邦”有其自身的局限，/ 但是，/ 布莱德劳博士指出 /
that anomaly detection is very tricky. / One may feel some symptoms / early / in the morning /
感知异常是件非常微妙的事。/ 有人可能会感觉到症状 / 很早 / 在凌晨 /
and have a heart attack / much later / in the afternoon. / Furthermore, /
而患心脏麻痹 / 很晚 / 在下午。/ 并且，/
Bluetooth instability, / smartphone viruses, / bugs and other faults, / and human error /
不稳定的蓝牙、/ 智能手机病毒、/Bug 和其他问题，/ 以及人类过失 /
are also potential risks. /
也是潜在的危险。/

Still, / the technology is advancing fast. / The future may be closer / than you think. /
仍然，/ 技术在飞速的进步。/ 未来可能更近 / 比你想的。/
Soon, / if you are stressed / too much / or your heartbeat is irregular, /
很快，/ 如果你压力 / 过多 / 或者心率不齐，/
your shirt may tell you / to take rest / or to seek medical help / immediately. /
你的衬衫可能会提示你 / 休息 / 或者寻求医疗救助 / 立即。/
Just be sure / to listen to your shirt's advice! /
切记 / 要遵从衬衫的建议！ /

经典语录之物联网篇

Notable Quotes on the Internet of Things

Sometimes it's useful to hear what industry leaders say about their field. It can give you some fresh insights. Here are some notable quotes on the Internet of Things.

"The Internet of Things, sometimes referred to as the Internet of Objects, will change everything—including ourselves."

— **Dave Evans**, chief futurist at Cisco

"If you think that the internet has changed your life, think again. The IoT is about to change it all over again!"

— **Brendan O'Brien**, co-founder and Chief Architect at Aria Systems

"The Internet of Things is not a concept; it is a network, the true technology-enabled Network of all networks."

— **Edewede Oriwoh**, professor at University of Bedfordshire

"Connecting products to the web will be the 21st century electrification"

— **Matt Webb**, CEO of BERG Cloud

"IoT will stump IT until clouds and big data come aboard."

— **Stephen Lawson**, EMC

"The IoT is big news because it ups the ante: 'Reach out and touch somebody' is becoming 'reach out and touch everything'."

— **Parker Trewin**, Senior Director of Content and Communications, Aria Systems

"As the next evolution of computing, the Internet of Things market will be bigger than all previous computing markets."

— **Greg Hodgson**, marketing director at Silicon Labs' Internet of Things solutions

"By 2018, 50% of the internet of things solutions will be provided by startups which are less than 3 years old."

— **Jim Tully**, research director at Gartner

"In order to move the Internet of Things forward, it's important to work together to define standards so that everyone can speak the same language."

— **Adam Justice**, VP of Grid Connect

"One of the myths about the Internet of Things is that companies have all the data they need, but their real challenge is making sense of it. In reality, the cost of collecting some kinds of data remains too high, the quality of the data isn't always good enough, and it remains difficult to integrate multiple data sources." — **Chris Murphy**, editor of Information Week

"The IoT is removing mundane repetitive tasks or creating things that just weren't possible before, enabling more people to do more rewarding tasks and leaving the machines to do the repetitive jobs."

— **Grant Notman**, head of Sales and Marketing at Wood & Douglas

"With emerging IoT technologies collecting terabytes of personal data the question is, are we ready to unbutton our online dress shirt while many are still just loosening their collars?"

— **Parker Trewin**, Senior Director of Content and Communications, Aria Systems

"When one system can connect thousands of devices and data streams across a rail network serving millions of people, that's the Internet of Things—and it's here right now."

— **Steve Pears**, talent managing director at Microsoft

"There's no stopping it. The Internet of things is coming, and you better disrupt or prepare to be disrupted." — **Joe Tucci**, CEO of EMC

核心语法

熟悉 # 关系代名词。# 关系代名词连接两个句子，兼具 # 连词和 # 代名词的作用。例文将介绍各种关系代名词。# 关系代名词有章可循，理解原理后学习将事半功倍。

- 50% of the solutions will be provided by **startups** (**which** are less than 3 years old). 50%的解决方案将被提供/由创业公司/成立不到3年的。
- Companies have **all the data** ([**that**] they need). 公司拥有所有数据/它们需要的。
- We'll hear **what** industry leaders say about their field. 我们将听到/业界领军人物对于他们从事领域的说法。
- **The Internet of Things**, [**which** is] sometimes referred to as the Internet of Objects, will change everything. 物联网，有时也称the Internet of Objects，将会颠覆一切。

熟悉 # 分词从句。# 分词从句是缩短句子的一种方式，可以省略那些通过推测即可得知的部分。理解 # 分词从句时首先要找出主语。# 分词从句有章可循，理解原理后学习将事半功倍。注意，第二个例句和第三个例句介绍了 # 分词从句和 # 关系代名词省略之间的差异。根据上下文理解即可。

- The IoT is doing many things, **enabling** more people to do more rewarding tasks and **leaving** the machines to do the repetitive jobs. IoT 在做很多事情，以让更多人做更有意义的事情，让机器做重复性的工作。
  ▶ enabling和 leaving的主语是The IoT。
- One system can connect thousands of devices across a rail network **serving** millions of people. 一个系统连接了数千台设备/遍及铁路网/为数百万人服务的。
  ▶ 分词从句的主语为“一个系统”。
- One system can connect thousands of devices across a rail network [which is] **serving** millions of people. 一个系统连接了数千台设备/遍及铁路网/正在为数百万人服务的。

▶ 如果视为关系代名词省略，那么工作的是“铁道网”或者“数千台设备”。根据上下文确定主语。

熟悉表从属关系的 #with 分词从句〈with+ 宾语 + 宾补〉。宾语和宾补是主谓关系，意为“做 ~ 的同时做着 ~”。

- **With** emerging IoT technologies **collecting** terabytes of personal data ~ 新兴物联网技术收集……的同时/太字节的个人数据……

熟悉 # 形式主语 it。真正的主语由于过长而后置。

- **It** remains difficult **to integrate** multiple data sources. 这仍然很难/综合众多数据资源。
- **It**'s important **to work** together to define standards. 这很重要/共同工作/定义标准。

单词&短语

useful 有用的
field 领域
insight 洞察力
notable 值得注意的
be referred to as ~ 被称为 ~
including ~ 包括 ~
be about to V 正准备做 ~
all over 全部，到处
concept 概念
connect A to B 将 A 和 B 联系起来
product 产品
electrification 电气化
stump 挑战
up the ante 提更多要求
reach out (to)(将手) 伸向
evolution 进化
previous 之前的
provide 提供
in order to V 目的是 ~
move forward 前进
define 定义
standard 标准
myth 神话
in reality 事实上
collect 收集
remain 保持
quality 质量
integrate 综合
multiple 众多的
data source 数据资源
emerge 出现
terabyte 太字节

be ready to V 准备做 ~
unbutton 解开纽扣
loosen 使松弛
rail network 铁路网
disrupt 破坏，使混乱
prepare to V 准备做 ~

根据提示完成句子

"The Internet of Things, /
"物联网，/

sometimes ________ __ __ the Internet of Objects, /
有时也称the Internet of Objects，/

will ______ __________ / - including ourselves." /
将改变一切/ 包括我们自己。" /

— **Dave Evans**, chief futurist at Cisco

"One of the ______ about the Internet of Things is /
"关于物联网的一个神话是/

that companies have all the data / ____ ____, /
公司拥有所有数据/ 他们需要的，/

but their ____ _________ is ________ _____ __ __. /
但他们真正的挑战是意识到这点。/

In reality, / the cost of collecting some kinds of data /
实际上，/ 收集各种数据的费用/

_______ ___ ____, / the quality of the data isn't always ____ ______, /
仍然很高，/ 数据的质量并不总是很好，/

and it _______ _________ / to _________ multiple data sources." /
并且仍然很难/ 综合众多数据。" /

— **Chris Murphy**, editor of Information Week

"The IoT is removing ________ ___________ ______ /

"物联网正在解除普通重复的任务/

or creating things / that just weren't possible before, /

或者创造事物/ 这在以前是不可能的，/

enabling more people to do more ___________ ______ /

以让更多的人从事更有意义的工作/

and leaving the machines to do the ___________ _____." /

并且让机器从事重复性工作。" /

— **Grant Notman**, head of Sales and Marketing at Wood & Douglas

"There's no stopping it. / The Internet of things is coming, /

"没有东西可以阻止。/ 物联网正在到来，/

and you _______ ________ or / prepare to be __________." /

你最好破坏它/否则就要准备好被破坏。" /

— **Joe Tucci**, CEO of EMC

思考题

讨论

1 请简述今后一两年内影响最大的物联网相关应用。

课堂总结

1 学习#关系代名词、#关系代名词省略的概念和例句。

2 学习#分词从句的构句方法、#with分词从句的概念和例句。

3 学习#形式主语、#真正主语的概念和例句。

翻译

Notable Quotes / on the Internet of Things

经典语录之物联网篇

Sometimes / it's useful / to hear / what industry leaders say / about their field. /
有时 / 这是有用的 / 听 / 业界领军人物讲述 / 他们从事的领域。/
It can give you / some fresh insights. / Here are some notable quotes / on the Internet of Things. /
这可以给你 / 一些新的启发。/ 以下为一些值得关注的名言 / 关于物联网的。/

"The Internet of Things, / sometimes referred to as the Internet of Objects, /
"物联网，/ 有时也称 the Internet of Objects，/
will change everything / - including ourselves." /
将改变一切 / 包括我们自己。"/

——**戴夫·埃文斯**，思科首席未来学家

"If you think / that the internet has changed your life, / think again. /
"如果你认为 / 网络会改变你的生活，/ 请三思。/
The IoT is about to change it / all over again!" /
IoT 将改变它 / 再一次完全地！"/

——**布兰登·奥－布莱恩**，Aria Systems 联合创始人、首席架构师

"The Internet of Things is not a concept; / it is a network, /
"物联网不是一个概念；/ 它是网络，/
the true technology-enabled Network of all networks." /
通过所有真正网络技术激活的网络。"/

——**艾德韦德·欧瑞沃**，贝德福德郡大学教授

"Connecting products to the web / will be the 21st century electrification." /
"将产品连接到网络 / 将会是 21 世纪的发电机。"/

——**马特·韦伯**，BERG Cloud 首席执行官

"IoT will stump IT / until clouds and big data come aboard." /
"物联网将挑战 IT/ 直到云和大数据加入。"/

——**史蒂芬·劳森**，易安信

"The IoT is big news / because it ups the ante: /
"物联网是大新闻 / 因为它提高了赌注：/
'Reach out and touch somebody' / is becoming / 'reach out and touch everything'." /
'伸出双手触碰所有人' 正在变为 / '伸出双手触碰所有事物'。"/

——**帕克·特里温**，Aria Systems 内容与通信部门高级总监

"As the next evolution of computing, / the Internet of Things market will be bigger /
"随着计算能力的再次飞跃 / 物联网市场将不断壮大 /
than all previous computing markets." /
相比之前的计算市场。"/

——**格雷戈·霍奇森**，硅谷实验室物联网解决方案营销总监

"By 2018, / 50% of the internet of things solutions will be provided / by startups /
"截止到 2018 年，/50% 的物联网解决方案将被提供 / 由创业公司 /
which are less than 3 years old." /
成立不到 3 年的。" /

——**吉姆·塔利**，Gartner 研究主管

"In order to move the Internet of Things forward, / it's important to work together /
"为了推动物联网前进，/ 团队合作很重要 /
to define standards / so that everyone can speak the same language." /
以定义标准 / 以便所有人可以使用同样的语言。" /

——**亚当·贾斯蒂斯**，Grid Connect 副总裁

"One of the myths about the Internet of Things is / that companies have all the data / they need, /
"关于物联网的一个神话是 / 公司拥有所有数据 / 他们需要的，/
but their real challenge is making sense of it. / In reality, / the cost of collecting some kinds of data /
但他们真正的挑战是意识到这点。/ 实际上，/ 收集各种数据的费用 /
remains too high, / the quality of the data isn't always good enough, / and it remains difficult /
仍然很高，/ 数据的质量并不总是很好，/ 并且仍然很难 /
to integrate multiple data sources." /
综合众多数据。" /

——**克里斯·墨菲**，《信息周刊》编辑

"The IoT is removing mundane repetitive tasks / or creating things /
"物联网正在解除普通重复的工作 / 或者创造事物 /
that just weren't possible before, / enabling more people to do more rewarding tasks /
在之前是不可能的，/ 以让更多的人从事更有意义的工作 /
and leaving the machines to do the repetitive jobs." /
让机器从事重复的工作。" /

——**格兰特·诺特曼**，Wood & Douglas 市场营销主管

"With emerging IoT technologies collecting / terabytes of personal data / the question is, /
"在新兴物联网技术收集……的同时 / 太字节的个人信息 / 问题在于，/
are we ready to unbutton our online dress shirt / while many are still just loosening their collars?" /
我们准备好解开在线礼服衬衫的纽扣了吗？ / 当很多人依然只是松了松他们的衣领。" /

——**帕克·特里温**，Aria Systems 内容与通信部门高级总监

"When one system can connect thousands of devices and data streams /
"当一个系统可以连接数千台的设备和数据流时 /
across a rail network / serving millions of people, / that's the Internet of Things / – and it's here right now."
遍及铁路网 / 为数百万人服务，/ 这就是物联网 /——它已经出现。" /

——**史蒂夫·比尔斯**，微软人才管理总监

"There's no stopping it. / The Internet of things is coming, /
"没有东西可以阻止它。/ 物联网正在到来，/
and you better disrupt or / prepare to be disrupted." /
你最好破坏它 / 否则就要准备好被破坏。" /

——**乔·图斯**，易安信首席执行官

第五部分

云

最近，亚马逊、微软、谷歌、甲骨文乃至IBM等几乎所有公司都在致力于云。云不再是一种流行语，而是一种常见的存在。本节将探讨云的概念和术语。

一切尽在云端！
Everything is Already in the Cloud!

These days you hear a lot about cloud computing. If you have no technical background, the term can be a bit intimidating. It sounds complicated, but most of us are already using the cloud [1]. Though you may not realize it, all your information is most likely in the cloud. How? Let's take a look.

Email

All web-based email services like Yahoo, Google (Gmail) and Microsoft (Hotmail) are cloud-based. Unless you run your own email server, you are using cloud-based services. That is, all of your contacts and emails are in the cloud.

Blogs & Websites

Blogs and websites such as Medium, Tumblr, Flickr, Instagram, and Pinterest are also cloud-based services. Again, all of your postings and photos are in the cloud.

Social Networking Sites

Facebook, Twitter, LinkedIn, and many other social media sites are hosted in the cloud. Your birthday, educational background, work history, and latest whereabouts are all in the cloud. Ah, your friends' information, too!

Mobile App Stores & Apps

Major mobile app stores, such as Google Play, Apple's App Store, and

Windows Marketplace, keep their apps and the account information in the cloud. If you buy apps from these major app stores, your purchase history and account information are in the cloud.

Gaming

Cloud gaming, according to an article in VentureBeat [2], could reach a turning point in 2015. Xbox Live, *World of Warcraft*, Steam and hundreds of game platforms are hosted in the cloud. Now with cloud-gaming capable devices, a massive number of players can interact with each other online. Most of these games store your game saves and comments in the cloud.

Productivity Tools

Productivity tools like word processors, spreadsheets, presentation programs, flowcharting applications, image editors, formula editors, and graph tools are available online as cloud-based services and applications. Google Docs and Microsoft Office 365 are good examples [3]. If you are using any of these tools, either at work or personally, your documents are stored in the cloud.

Online Storage

Nowadays most people own multiple devices—from laptops to smart-phones and tablets—and they want to access their data from anywhere, at any time, from any connected device. Online storage services are used for that reason, and Microsoft OneDrive, Apple iCloud Drive, Google Drive, and Dropbox are popular ones [4]. Needless to say, these services store your data in the cloud.

出 处

1. Typical consumer cloud computing scenarios are you using the cloud? http://goo.gl/gEjxHP
2. Cloud gaming could reach a turning point in 2015, http://goo.gl/MyqSFP
3. Battle of the Mobile Office Suites: Microsoft Office vs. Google Docs, http://goo.gl/in14dA
4. OneDrive, Dropbox, Google Drive, and Box: Which cloud storage service is right for you? http://cnet.co/12eQ8HW

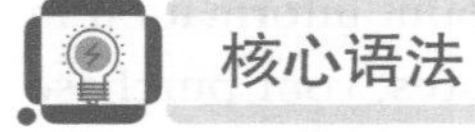

核心语法

熟悉 # 句式 2〈主语 + 动词 + 补语（名词、形容词）〉。使用 #be 动词和 # 感官动词 sound。#be 动词意为“是 ~”，# 感官动词 sound 意为“听起来 ~”。

- The term **is** intimidating. 这个术语是令人生畏的（=使人发憷）。
- Many websites **are** cloud–based services. 很多网站是基于云的服务。
- It **sounds** complicated. 这听起来很复杂。

熟悉 # 句式 3〈主语 + 动词 + 宾语〉。使用 # 名词和 # 不定式 to 作 # 宾语。# 不定式 to 的 # 名词性用法意为“做 ~”。want 后可接 # 不定式 to 作 # 宾语。

- Most of us are using the cloud. 我们中的大多数都在使用云。
- Most people own multiple devices. 多数人拥有众多设备。
- They want **to access** their data. 他们想/访问他们的数据。

熟悉 # 被动语态〈be+ 过去分词〉。意为“被 ~”。

- Many sites **are hosted** in the cloud. 很多站点被托管到云端。
- Your documents **are stored** in the cloud. 你的文件被保存到云。
- Online services **are used** for that reason. 线上服务因为那个原因被使用。

使用 # 并列句介绍列举示例的各种方式。连接多个相同的句子成分时，在最后一项前加 # 并列连词 and、or、but 等。

- All web–based email services like Yahoo, Google **and** Microsoft are cloud-based. 雅虎、

谷歌、微软之类的网页端邮箱服务都基于云。

- Blogs and websites such as Medium, Tumblr, Flickr, Instagram, **and** Pinterest are also cloud–based services. Medium、Tumblr、Flickr、Instagram和Pinterest等博客和网站也是基于云的服务。
- Facebook, Twitter, LinkedIn, **and** many other social media sites are hosted in the cloud Facebook、Twitter、LinkedIn和其他众多社交媒体网站都被托管至云。
- Nowadays most people own multiple devices　from laptops to smartphones **and** tablets. 现在大多数人拥有众多设备——从台式电脑到智能手机和平板电脑。

熟悉连接两个句子的 # 连词。

- **If** you have no technical background, ~ 如果你没有技术背景，~
- **Though** you may not realize it, ~ 即使你可能没有意识到，~
- **Unless** you run your own server, ~ 除非你操作自己的服务器，~

单词&短语

these days 最近，近来
technical background 技术背景
term 术语
intimidating 令人生畏的
complicated 复杂的
realize 意识到
likely ~ 可能 ~
–based 基于 ~
that is 即，换言之
contacts 联系方式
educational background 学历
work history 工作经历
whereabouts 现状，所在，踪迹
account information 账户信息
purchase history 购买历史
according to ~ 依据 ~，按照 ~
reach a turning point 到达转折点
hundreds of ~ 数百个 ~
capable 能胜任的
a massive number of ~ 大量的
interact with ~ 与 ~ 互动
game save（游戏）存档
productivity tools 生产工具
formula editor 公式编辑器
personally 个人地
needless to say 无需赘言

技术术语

multiple devices 多设备指 PC、智能手机、平板电脑、游戏主机、智能 TV 等个人所有且同时使用的众多设备。

根据提示完成句子

______ ______ / you hear a lot / _____ cloud computing. /
最近/ 你听到很多言论/ 关于云计算。/

__ you have no ________ _________, / the ____ can be a bit __________. /
如果你没有技术背景，/ 这个术语会有些让人生畏。/

It ______ _________, / but most of us ___ already _____ the cloud. /
它（术语）听起来很复杂，/ 但我们中的大多数已经在使用云。/

______ you may not realize it, / ___ ____ __________ is most likely /
即使你可能没有意识到这点，/ 但你的所有信息可能已经/

__ ___ _____.
在云端。/

How? / _____ take a look. /
这是怎么回事呢？/让我们一起看一下。/

______ _______ /
网络硬盘

Nowadays ____ ______ ___ ________ _______ /
近来很多人群使用多设备/

- ____ laptops __ smartphones and tablets - /
——从台式机到智能手机和平板电脑——/

and ____ ____ __ ______ _____ ____ / from anywhere, / __ ___ ____, /
他们想访问自己的数据/随地，/ 随时，/

____ ___ _________ ______. /
在任何连接设备上。/

思考题

解答题

1 （理解）以下哪一项不属于云服务？

ⓐ 存储文件　　ⓑ 社交网络

ⓒ 应用商城　　ⓓ 存储个人公证书

2 （理解）以下哪项与其他各项不同？

ⓐ Google Play　　ⓑ Microsoft OneDrive

ⓒ Dropbox　　ⓓ Apple iCloud

3 （论述）简述如何借助计算能力——而非软件——使用云？

4 （论述）简述云文件共享服务如何实现多设备同步。会因没有同步而出现困境吗？该如何应对？

讨论

1 请简述云使用过程中侵犯的个人隐私（信息）及应对方法。

2 随着云的发展，本地硬盘和USB存储器会完全消失吗？

课堂总结

1 学习#句式2中使用的动词和例句。

2 学习使用#不定式to作#宾语的动词和例句。

3 学习#过去分词、#被动语态、#主动语态的概念。

答案

1 d.（向云上传个人敏感信息时可能出现问题。） 2 a.（只有a是应用商城，其他都是在线存储方式。） 3 使用IaaS。AWS EC2和DigitalOcean等都是提供计算能力的代表性服务。 4 在同步过程中使用日期和文件散列值监控冲突（最新变更的内容优先级更高）。如果同步失败，而提供显示文件更新记录的备份功能，则可以使用能重回需要日期的功能（point-in-time recovery）进行复原。

翻译

Everything is Already in the Cloud!

一切尽在云端!

These days / you hear a lot / about cloud computing. / If you have no technical background, /
最近 / 你听到很多言论 / 关于云计算。/ 如果你没有技术背景，/
the term can be a bit intimidating. / It sounds complicated, / but most of us are already using the cloud. /
这个术语会有些让人生畏。/ 它听起来很复杂，/ 但我们中的大多数已经在使用云。/
Though you may not realize it, / all your information is most likely / in the cloud. /
即使你可能没有意识到这点，/ 但你的所有信息可能已经 / 在云端。/
How? / Let's take a look. /
这是怎么回事呢？ / 让我们一起看一下。/

Email / 邮件

All web-based email services / like Yahoo, Google (Gmail) and Microsoft (Hotmail) / are cloud-based. /
所有网页端邮箱 / 例如雅虎、谷歌（Gmail）、微软（Hotmail）/ 都基于云。/
Unless you run your own email server, / you are using cloud-based services. /
除非你运行自己的邮箱服务器，/ 否则你就在使用云服务。/
That is, / all of your contacts and emails are / in the cloud. /
换言之，/ 所有联系方式和邮件 / 都在云。/

Blogs & Websites / 博客和网站

Blogs and websites / such as Medium, Tumblr, Flickr, Instagram, and Pinterest /
博客和网站 / 例如 Medium、Tumblr、Flickr、Instagram and Pinterest/
are also cloud-based services. / Again, all of your postings and photos are / in the cloud. /
也是云服务。/ 同样，你的所有帖文和照片 / 都在云。/

Social Networking Sites / 社交网站

Facebook, Twitter, LinkedIn, and many other social media sites are hosted / in the cloud. /
Facebook、Twitter、领英和其他社交媒体网站被托管 / 到云端。/
Your birthday, educational background, work history, and latest whereabouts are all / in the cloud. /
你的生日、教育背景、工作经历和近况全部都 / 在云端。/
Ah, your friends' information, too! /
啊，也包括你朋友的信息！ /

Mobile App Stores & Apps / 手机应用商城和应用

Major mobile app stores, / such as Google Play, Apple's App Store, and Windows Marketplace, /
主要的手机应用商城，/ 例如 Google Play、苹果的 App Store 和 Windows Marketplace，/
keep their apps and the account information / in the cloud. /
将它们的应用和账户信息 / 放在云端。/

If you buy apps from these major app stores, /
如果你在这些主要的应用商城购买应用，/
your purchase history and account information are / in the cloud. /
你的购买记录和账户信息 / 就在云内。/

Gaming / 游戏

Cloud gaming, / according to an article in VentureBeat [2], / could reach a turning point in 2015. /
云游戏，/ 根据 VentureBeat 的报道，/ 在 2015 年达到转折点。/
Xbox Live, World of Warcraft, Steam and hundreds of game platforms / are hosted / in the cloud. /
Xbox Live、魔兽世界、Steam 和数百个游戏平台 / 被托管到 / 云。/
Now with cloud-gaming capable devices, /
现在使用云游戏存储设备，/
a massive number of players can interact with each other online. /
众多玩家可以与其他玩家在线互动。/
Most of these games store your game saves and comments / in the cloud. /
多数游戏将你的存档和评论 / 放在云端。/

Productivity Tools / 生产工具

Productivity tools / like word processors, spreadsheets, presentation programs,
生产工具 / 例如文字处理软件、电子数据表、PPT、/
flowcharting applications, image editors, formula editors, and graph tools /
流程图应用、图片编辑器、公式编辑器和图表工具 /
are available online / as cloud-based services and applications. /
可以在线使用 / 作为云服务和应用。/
Google Docs and Microsoft Office 365 are good examples [3]. / If you are using any of these tools, /
谷歌 Docs 和 Microsoft Office 365 就是很好的例证。/ 如果你正在使用其中任何一种工具，/
either at work or personally, / your documents are stored / in the cloud. /
在工作中或个人使用，/ 你的文档被保存 / 到云。/

Online Storage / 网络硬盘

Nowadays most people own multiple devices / - from laptops to smartphones and tablets - /
近来很多人群使用多设备 /——从台式机到智能手机和平板电脑——/
and they want to access their data / from anywhere, / at any time, / from any connected device. /
他们想访问自己的数据 / 随地，/ 随时，/ 在任何连接设备上。/
Online storage services are used / for that reason, /
网络云盘被使用的 / 原因就在此，/
and Microsoft OneDrive, Apple iCloud Drive, Google Drive, and Dropbox are popular ones [4]. /
Microsoft OneDrive、Apple iCloud Drive、Google Drive 和 Dropbox 比较流行。/
Needless to say, / these services store your data / in the cloud. /
无需赘言，/ 这些服务保存你的数据 / 到云端。/

向非技术圈朋友解释云

How to Explain the Cloud to Non-Techie Friends

Your friend has no technical background, yet she is curious about technology. Recently she asked you what Cloud Computing is. What would you tell her? How would you explain its concept and various models to someone with little or no technical background? Here's a suggestion [1]. Compare Cloud Computing to everyone's favorite food, pizza!

That's exactly what HP does. In a YouTube video titled *Cloud Pizza*, HP Norway's Technology Director, Stig Alstedt, explains Cloud Computing by comparing it with pizza [2]. Let's see how.

Homemade Pizza = Traditional IT

Back in the old days, if you wanted pizza for dinner, you had to make your own. In that case, you buy all the ingredients, knead the dough, make the crust, chop the vegetables, and bake the pizza. You also need proper cookware. The process requires expertise, but you get exactly what you want, how you want it, when you want it. Homemade Pizza is equal to building your own IT infrastructure with your own resources.

Frozen Pizza = Private Cloud

Nowadays supermarkets offer customers a variety of frozen pizzas. Pizza manufacturers do almost all the work. You just select one from the frozen pizza aisle, bring it home, and bake it in your own oven. It's convenient, but the flavors and prices are fixed. Frozen Pizza is similar to Private Cloud.

Delivered Pizza = Managed Cloud

Delivered pizza is a bit more convenient than frozen pizza. You order what you want and how you want, then the pizza is delivered to your doorstep. No baking is necessary. However, menu items are limited and you still need your own dishes. Delivered Pizza is close to Managed Cloud.

Restaurant Pizza = Public Cloud

Restaurant pizza requires the least expertise because the restaurant provides you with everything. You simply show up, order, eat and pay. You get one invoice at the end of your meal. However, other customers also dine in, so the service can be slow when the place is crowded. Furthermore, everyone orders from the same menu. It means there is not much room for individual customization. Restaurant Pizza is comparable to Public Cloud, except that Public Cloud is often the most inexpensive option while restaurant pizza is not.

Your Own Topping = Hybrid Cloud

Many pizza delivery places and restaurants offer an option to choose your own topping. Hybrid Cloud is like choosing your own toppings to make your favorite pizza.

出 处

1. 10 Interesting Cloud Computing Analogies, http://goo.gl/P4yGsF
2. CloudPizza, https://goo.gl/OPougl

核心语法

熟悉 # 句式 4〈主语 + 动词 + 间接宾语 + 直接宾语〉。# 句式 4 中的动词意为“为 ~ 做 ~”。例文也将同时介绍易与 # 句式 4 的动词混淆的 # 句式 3 中的动词。

- Supermarkets offer customers a variety of frozen pizzas. 超市提供/给顾客/各种各样的冷冻披萨。
- You explain its concept to your friend. 你向你的朋友介绍它的概念。
  You explain your friend its concept. (×)
- The restaurant provides you with everything. 餐厅向你提供所有东西。
  The restaurant provides you everything. (×)

熟悉 # 关系代名词 what。# 关系代名词 what 等同于 the thing(s) that。所以包含 # 先行词，意为“~ 做的~”。注意，以下最后一个例句中的 what 为 # 疑问词，意为“什么”。语法上比较难区分，但无需拘泥于此，根据上下文理解即可。

- That's exactly **the thing** + HP does **it**. 那就是这件事 + 惠普做的。
  = That's exactly **the thing** (**that** HP does). =那就是这件事/惠普做的。
  = That's exactly **what** HP does. =那就是惠普做的这件事。
- You order **the thing** + You want **it**. 你订购了这个东西 + 你想要的。
  = You order **the thing** (**that** you want). =你订购了这个东西/你想要的。
  = You order **what** you want. =你订购了/你想要的东西。
- She asked you **what** Cloud Computing is. 她问你/什么是云计算。

从句是导致句子复杂化的主要原因。多个从句放在一起组成复句。根据整体句意理解从句即可。例文将介绍 #that 从句。

- It means [that] **there is no room for individual customization**. 这意味着/没有空间/为个人定制的。
- Restaurant Pizza is comparable to Public Cloud, except **that Public Cloud is often the most inexpensive option**. 堂食披萨可以与公共云相媲美/除了/公共云是最低廉的选择。

单词&短语

explain A to B 向 B 解释 A
technical background 技术背景
be curious about ~ 对 ~ 很好奇
concept 概念
suggestion 建议
compare A to/with B 将 A 和 B 比较
exactly 确切地
back in the old days 过去
ingredient 材料，成分
knead 揉，捏（面团）
chop 切碎
proper 合适的
cookware 烹饪工具
require 要求，需要
expertise 专业知识技术
be equal to ~ 等同于 ~
resource 资源
nowadays 最近（=these days）
a variety of 各种各样的（=various）
manufacturer 制造商
aisle 过道
convenient 方便的
flavor 风味，味道
frozen pizza 冷冻披萨
be similar to ~ 类似于 ~
delivered 寄送
order 订购
at one's doorstep 门前
necessary 必须的
limited 有限的
be close to ~ 接近于 ~
least 至少
provide A with B 向 A 提供 B
show up 到场
invoice 发票
dine in 就餐
crowded 拥挤的
furthermore 而且
be comparable to ~ 和 ~ 相媲美
except (that) ~ 除了 ~
inexpensive 低廉的
choose 选择

根据提示完成句子

Restaurant pizza requires the ____ __________ /
堂食披萨对专业性要求最少/

because the restaurant provides you / with everything. /
因为餐厅给你提供/ 一切东西。/

You simply ____ __, / _____, / ___ and ___. / You ___ ___ _______ /
你仅需要到场、/ 下单、/ 食用然后付钱。/ 你拿到一张发票/

__ ___ ___ __ of your meal. / However, / other customers also dine in, /
在就餐结束时。/ 但是，/ 其他顾客也会就餐，/

so the service ___ __ ____ / when the _____ is _______. / Furthermore, /
所以服务会缓慢/ 当餐厅拥挤时。/ 而且，/

everyone orders from the same menu. / It means /
每人都通过同一个菜单下单。/ 这意味着/

there is not ____ ____ / for __________ ____________. /
没有余地/ 为个人定制。/

Restaurant Pizza __ __________ to Public Cloud, / except /
堂食披萨可以和公共云相媲美，/ 除了/

that Public Cloud is often the ____ ___________ option /
公共云通常是最廉价的选择/

while restaurant pizza is not. /
而堂食披萨并不是。/

思考题

解答题

1 （理解）以下哪项与外卖披萨最相似？

ⓐ 传统 IT　　ⓑ 私有云

ⓒ 公共云　　ⓓ 社区云

2 （理解）设备投资和基础技术最需要的是以下哪一种云？

ⓐ 传统 IT　　ⓑ 私有云

ⓒ 公共云　　ⓓ 混合云

3 （论述）公共云是快速获取云服务的不二之选，在保证这种优势的前提下使用公共云时，可采取何种方法降低使用成本？

4 （论述）简述堂食披萨和公共云的最大区别。

讨论

1 假设有一家公司需要快速提供海外服务，选择哪种云服务比较有利？ 请阐述理由。

2 调查并以菜单形式总结知名云服务（亚马逊AWS、微软Azure、Google Cloud）。

课堂总结

1 学习#易与句式4动词混淆的句式3中的动词。

2 学习#关系代名词的概念和例句。

3 学习#名词从句和#形容词从句的例句。

答案

1 d. 2 a. 3 签订长期合同。以竞标形式购买较便宜时间段的计算能力。 4 公共云的成本比堂食披萨低。

翻译

How to Explain the Cloud / to Non-Techie Friends

向非技术圈朋友解释云

Your friend has no technical background, / yet she is curious / about technology. /
你的朋友没有技术背景，/ 但是她很好奇 / 技术。/
Recently / she asked you / what Cloud Computing is. / What would you tell her? /
最近 / 她问你 / 什么是云计算。/ 你将告诉她什么？ /
How would you explain its concept and various models / to someone /
你将如何说明概念和各种模式？ / 给那些 /
with little or no technical background? /
有很少或者没有技术背景的人 /
Here's a suggestion [1]. Compare Cloud Computing / to everyone's favorite food, / pizza! /
这里有一个建议。/ 将云计算比喻为 / 大家最喜爱的美食 / 披萨！ /

That's exactly what HP does. / In a YouTube video / titled Cloud Pizza, /
准确来说这就是惠普所做的。/ 在一个 YouTube 视频 / 名为云披萨，/
HP Norway's Technology Director, /
惠普挪威技术总监 /
Stig Alstedt, / explains Cloud Computing / by comparing it / with pizza [2]. / Let's see how. /
斯蒂格 · 埃尔斯泰德，/ 解释云计算 / 通过将其和披萨做比较。/ 让我们一探究竟。/

Homemade Pizza = Traditional IT / 自制披萨 = 传统IT

Back in the old days, / if you wanted pizza / for dinner, / you had to make your own. / In that case, /
过去，/ 如果你想吃披萨 / 当晚饭，/ 你必须自己做。/ 这种情况下，/
you buy all the ingredients, / knead the dough, / make the crust, / chop the vegetables, /
你购买所有材料，/ 和面、/ 做披萨皮、/ 切菜、/
and bake the pizza. / You also need proper cookware. / The process requires expertise, /
然后烤披萨。/ 你也需要合适的烹饪工具。/ 这个过程需要专业知识，/
but you get exactly / what you want, / how you want it, / when you want it. /
但你确实得到了 / 你想要的，/ 以你想要的方式，/ 在你想要的时候。/
Homemade Pizza is equal to building your own IT infrastructure / with your own resources. /
自制披萨就等同于构建你自己的 IT 基础设施 / 用你自己的资源。/

Frozen Pizza = Private Cloud / 冷冻披萨= 私有云

Nowadays / supermarkets offer / customers / a variety of frozen pizzas. /
近来 / 超市出售给 / 顾客 / 各种冷冻披萨。/

Pizza manufacturers do almost all the work. /
披萨制造商几乎承担了所有工作。/
You just select one / from the frozen pizza aisle, / bring it home, / and bake it / in your own oven. /
你仅需要选择一个 / 从冷冻披萨区，/ 带回家，/ 然后自己烤 / 用你自己的微波炉。/
It's convenient, / but the flavors and prices are fixed. / Frozen Pizza is similar to Private Cloud. /
这是方便的，/ 但味道和价格是固定的。/ 冷冻披萨类似于私有云。/

Delivered Pizza = Managed Cloud / 外卖披萨 = 社区云

Delivered pizza is a bit more convenient / than frozen pizza. / You order / what you want /
外卖披萨比较方便 / 相比冷冻披萨。/ 你下单 / 你想要的 /
and how you want, / then the pizza is delivered / to your doorstep. / No baking is necessary. /
并且以你喜欢的方式，/ 然后披萨被送到 / 你的门前。/ 无需再烤。/
However, menu items are limited / and you still need your own dishes. /
但菜单项目有限 / 并且你仍然需要用你自己的盘子。/
Delivered Pizza is close to Managed Cloud. /
外卖披萨接近社区云。/

Restaurant Pizza = Public Cloud / 堂食披萨 = 公共云

Restaurant pizza requires the least expertise / because the restaurant provides you / with everything. /
堂食披萨对专业性的要求最少 / 因为餐厅给你提供 / 一切东西。/
You simply show up, / order, / eat and pay. / You get one invoice / at the end of your meal. / However, /
你仅需要到场、/ 下单、/ 食用并付钱。/ 你拿到一张发票 / 就餐结束时。/ 但是，/
other customers also dine in, / so the service can be slow / when the place is crowded. / Furthermore, /
其他顾客也会就餐，/ 所以服务速度会很慢 / 当餐厅拥挤时。/ 而且，/
everyone orders from the same menu. / It means / there is not much room / for individual customization. /
每个人通过同一个菜单下单。/ 这意味着 / 没有空间 / 为个人定制。/
Restaurant Pizza is comparable to Public Cloud, / except /
堂食披萨可以和公共云相媲美，/ 除了 /
that Public Cloud is often the most inexpensive option / while restaurant pizza is not. /
公共云通常是最廉价的选择 / 而堂食披萨并不是。/

Your Own Topping = Hybrid Cloud / 自选配料 =混合云

Many pizza delivery places and restaurants offer an option / to choose your own topping. /
很多披萨外卖店和餐厅提供选择 / 让你亲自选择配料。/
Hybrid Cloud is like / choosing your own toppings / to make your favorite pizza. /
混合云就类似于 / 选择自己的配料 / 制作你最喜爱的披萨。/

数值中反映的未来

Numbers Tell the Future

Numbers can give a sense of the future landscape of the IT industry. Here are some numbers that every technical executive should consider when evaluating his or her company's cloud plan.

- The **global cloud services market** is estimated to increase from $209.9 billion in 2014 to $555 billion in 2020, growing at a compound annual growth rate (CAGR) of 17.6% [1].

- The **cloud equipment market** is predicted to reach $79.1 billion by 2018. Cloud equipment includes servers, storage, networking hardware and high-speed links [2].

- IDC forecasts **public cloud services spending** will increase at a compound annual growth rate (CAGR) of 22.8% to $127.5 billion by 2018, including $82.7 billion for SaaS, $24.6 billion for IaaS, and $20.3 billion for PaaS [3].

- It is forecasted that **global SaaS software revenues** will reach $106 billion in 2016. By 2018, SaaS (Software at a Service) workloads will take up 59% of the total cloud workloads, which is up from 41% in 2013 [4].

- More than half of the US government has already moved to the cloud, and federal agencies are continuing to invest in cloud computing. **Federal cloud services spending** is expected to grow from $2.3 billion in 2013 to $6.1 billion in 2018 [5].

- IT decision makers will increase their spending on cloud computing by 42% in 2015, says **Computerworld Forecast Study 2015** [6].

- The **2014 Future of Cloud Computing survey** of 1,358 respondents including users and vendors reveals 49% are using cloud solutions for revenue generation or product development and 45% are already running or plan to run their company from the cloud [7].

- The **2014 IDG Enterprise Cloud Computing Survey** of 1,672 IT and security decision-makers finds that more than two-thirds (69%) of companies already have at least one application in the cloud. The remaining companies plan to use cloud solutions within the next three years [8].

Three main drivers for cloud investments are increasing speed of deployment, lowering total cost of ownership (TCO), and replacing legacy technology. Average cloud investment in 2014 is $1.6 million, up 19% from 2012. The survey shows enterprises with more than 1,000 employees invest significantly more than small businesses.

Cloud security and privacy still remain top concern for businesses. A majority of respondents (56%) say their number one challenge for migrating to the cloud is an uncertain ability to enforce their security policies at cloud provider sites [8].

出 处

1. Global Cloud Services Market is Expected to Reach $555 Billion, Globally, by 2020, http://goo.gl/1pOHyr
2. Infographic : Cloud's Growing Footprint in Storage, http://goo.gl/ma9Fls
3. Forecasts Call For Cloud Burst Through 2018, http://goo.gl/x9yeY3
4. Cisco Global Cloud Index: Forecast and Methodology, 2013–2018, http://goo.gl/uyGTjK
5. Federal Agencies Making Gains in Cloud and Data Center Consolidation Despite Budget Constraints, According to Deltek, http://goo.gl/6U882J
6. Computerworld Forecast Study 2015, http://goo.gl/xi9LqO
7. 2014 Future of Cloud Computing 4th Annual Survey Results, http://goo.gl/F3nS2T
8. IDG Enterprise Cloud Computing Study 2014, http://goo.gl/k69clp

核心语法

熟悉 # 分词从句。# 分词从句遵循一定的规则，是缩短句子的方法之一。具体而言，# 状语从句〈连词 + 主语 + 动词〉中去掉可充分推测的部分，并将动词变为分词。例文介绍的分词从句位于句子后半部分。

- Every CTO should consider these numbers **when evaluating** a cloud plan. 每个首席技术官应该考虑这些数值/评估云计划时。
  = Every CTO should consider these numbers **when he or she evaluates** a cloud plan. =每个首席技术官应该考虑这些数值/他（她）评估云计划时。
- The market will increase from $209 billion to $555 billion, **growing** at a rate of 17.6% per year. 市场将会增长/从2090亿美元到5550亿美元，年增长率为17.6%。
  = The market will increase from $209 billion to $555 billion, **as it grows** at a rate of 17.6% per year. =市场将会增长/从2090亿美元到5550亿美元，（市场的）年增长率为17.6%。

熟悉 # 关系代名词的省略。# 宾格关系代名词、# 主格关系代名词 +be 动词可以省略。省略后不影响原意。

- Here are **some numbers** + Every CTO should consider **them**. 这些数值 + 所有首席技术官都应该考虑它们。
  = Here are **some numbers** ([**that**] every CTO should consider). =这些数值/所有首席技术官都应该考虑的。
- SaaS workloads will take up **59%** of the total cloud workloads, **and it** is up from 41% in 2013. SaaS工作负载将占据整个云工作负载的59%，从2013年的41%上升至此。
  = SaaS workloads will take up **59%** of the total cloud workloads, [**which** is] up from 41% in 2013. =SaaS工作负载占整个云工作负载的59%，从2013年的41%上升至此。

熟悉新闻报道中常见的推测表达。

- The market is estimated to increase from $209.9 billion in 2014 to $555 billion in 2020. 市场预计会增长/从2014年的2099亿美元/到2020年的5550亿美元。
- The spending is expected to grow from $2.3 billion in 2013 to $6.1 billion in 2018. 支出预计会增长/从2013年的23亿美元/到2018年61亿美元。
- The market is predicted to reach $79.1 billion by 2018. 市场预计将达到791亿美元/到2018年。
- The spending will increase at a rate of 22.8% to $127 billion by 2018. 支出将增长/以22.8%的增长率/截止到2018年/到127亿美元。
- It's forecasted that the revenues will reach $106 billion in 2016. 预计/利润将达到106亿美元/到2016年。

单词&短语

sense 感觉
landscape 风景
executive 总经理
consider 考虑
evaluate 评估
be estimated to V 估计为 ~
increase 增加
compound annual growth rate
年均复合增长率（CAGR）
cloud equipment 云设备
be predicted to V 被预测为 ~
reach 达到
forecast 预测
spending 支出
including ~ 包括 ~
revenue 利润
workload 工作负载
federal agency 联邦机构
continue to V 持续做 ~
invest in 在 ~ 上投资
be expected to V 被期望做 ~
decision maker 决策者
respondent 应答者
vendor 供应商
reveal 揭露
revenue generation 创收
product development 产品研发
remaining 剩余的
driver 推动者
deployment 部署，调度
total cost of ownership 总拥有成本（TCO）
replace 替换

技术术语

IaaS（Infrastructure as a Service）基础设施即服务提供计算机

（物理机器和虚拟机器）。

PaaS（Platform as a Service）平台即服务提供计算平台，包括操作系统、编程语言运行环境（栈）、数据库、Web 服务器。

SaaS（Software as a Service）软件即服务向用户提供应用程序软件和数据库。

根据提示完成句子

Three ____ ______ / for _____ ________ /
三大主要原因/ 面向云投资的/

are increasing _____ __ ___________, /
是提高部署速度，/

lowering _____ ____ __ _________ (TCO), /
降低总拥有成本（TCO），/

and replacing ______ ____________. / Average _____ __________ /
和取代传统技术。/ 平均云投资/

in 2014 / is $1.6 million, / up 19% from 2012. / The ______ _____ /
在2014年/ 是1600万美元，/ 2012年提高了19%。/ 调查显示/

enterprises with more than 1,000 employees /
员工超过1000人的公司/

invest significantly more than small businesses. /
会比小型企业投资更多。/

Cloud ________ and _______ still ______ ___ _______ for businesses. /
云安全和隐私仍然是业务中考虑最多的要素。/

A _________ of ___________ (56%) say their ______ ___ __________
多数应答者（56%）提到他们最大的忧虑/

for _________ / to the cloud / is an _________ _______ /
关于迁移/ 到云/ 是一种不确定的能力/

to ______ their ________ ________ / at _____ _________ sites [8]. /

将执行安全政策的/　　　　　　　　　　在云供应商网站。/

思考题

解答题

1 （理解）以下哪项不属于云设备？

ⓐ 服务器　　　　ⓑ 监控器

ⓒ 快速网络开关　　ⓓ 内存

2 （论述）请简述小企业引入云时最需要考虑的事项。

讨论

1 亚马逊发布2015年第一季度销售业绩时首次公开了公司的销售额（15亿7000万美元）。以此预估整体IaaS云市场2015年的销售额。

2 尽管服务器价格下降，但研发/运营人员却不断加速使用云，根本原因何在？

3 请用可视化方法总结例文给出的数值，以帮助那些云引入过程中需要决策的人。

课堂总结

1 学习#分词从句的概念和#独立分词从句。

2 学习#关系代名词的概念和例句。

3 学习#准关系代名词的概念和例句。

答案

1 b.（云服务器多是虚拟的，所以多数没有监控器。） 2 安全和个人信息保护问题

翻译

Numbers Tell the Future

数值中反映的未来

Numbers can give a sense of the future landscape / of the IT industry. / Here are some numbers /
数值能够提供对未来风景的感知 /IT 业界的。/ 这里有很多数值 /
that every technical executive should consider / when evaluating his or her company's cloud plan. /
所有首席技术官应当考虑的 / 评估他（她）公司的云计划时。/

- The global cloud services market is estimated / to increase / from $209.9 billion in 2014 /
  全球云服务市场预计/增长/从2014年的2099亿美元/
  to $555 billion in 2020, / growing at a compound annual growth rate (CAGR) of 17.6% [1]. /
  到2020年的5550亿美元，/以17.6%的年增长率。/

- The cloud equipment market is predicted / to reach $79.1 billion / by 2018. /
  云设备市场预计/达到791亿美元/到2018年。/
  Cloud equipment includes servers, storage, networking hardware and high-speed links [2]. /
  云设备包括服务器、内存、网络硬件和高速链接。/

- IDC forecasts / public cloud services spending will increase /
  IDC预测/公共云服务支出将会增加/
  at a compound annual growth rate (CAGR) of 22.8% / to $127.5 billion / at 2018, /
  以22.8%的年增长率/至1275亿美元到2018年，/
  including $82.7 billion / for SaaS, / $24.6 billion / for IaaS, / and $20.3 billion for PaaS [3]. /
  包括827亿美元/面向SaaS的，/246亿美元/面向IaaS的，/和203亿美元/关于PaaS的。/

- It is forecasted / that global SaaS software revenues will reach $106 billion / in 2016. / By 2018, /
  预计/全球SaaS软件销售额将达到106亿美元/在2016年。/截止到2018年，/
  SaaS (Software at a Service) workloads will take up 59% of the total cloud workloads, /
  SaaS工作负载将占云总工作负载的59%，/
  which is up / from 41% / in 2013 [4]. /
  增长/从41%/2013年的。/

- More than half of the US government has already moved to the cloud, /
  一半以上的美国政府业务已经迁移至云，/
  and federal agencies are continuing / to invest / in cloud computing. /
  联邦政府机关也继续/投资/云计算。/

Federal cloud services spending is expected / to grow / from $2.3 billion / in 2013 /
联邦云服务支出预计/会增长/从23亿美元/在2013年/
to $6.1 billion / in 2018 [5]. /
到61亿美元/在2018年。/

- IT decision makers will increase their spending / on cloud computing / by 42% in 2015, /
IT决策者将提高支出/在云计算上/至42%/在2015年，/
says Computerworld Forecast Study 2015 [6]. /
据“《计算机世界》预测研究2015”称。/

- The 2014 Future of Cloud Computing survey of 1,358 respondents / including users and vendors /
1358人参与的2014未来云计算调查/包括用户和供货商/
reveals / 49% are using cloud solutions / for revenue generation / or product development /
显示/49%在使用云解决方案/为创收/或者产品研发/
and 45% are already running / or plan to run their company / from the cloud [7]. /
45%已经在运营/或者计划运营公司/通过云。/

- The 2014 IDG Enterprise Cloud Computing Survey of 1,672 IT and security decision-makers finds /
1672名IT和安全决策人士参与的2014 IDG大企业云计算调查发现/
that more than two-thirds (69%) of companies already have at least one application / in the cloud. /
2/3以上（69%）的公司至少已经拥有一款应用/在云。/
The remaining companies plan / to use cloud solutions / within the next three years [8]. /
其他公司计划/使用云解决方案/在未来3年内。/

Three main drivers / for cloud investments / are increasing speed of deployment, /
三大主要原因 / 面向云投资的 / 是提高部署速度，/
lowering total cost of ownership (TCO), /
降低总拥有成本（TCO），/
and replacing legacy technology. / Average cloud investment / in 2014 / is $1.6 million, /
和取代传统技术。/ 平均云投资 / 在 2014 年 / 为 1600 万美元，/
up 19% from 2012. / The survey shows / enterprises with more than 1,000 employees /
从 2012 年的 19% 增长至此。/ 调查显示 / 员工超过 1000 人的公司 /
invest significantly more than small businesses. /
会比小型企业投资更多。/

Cloud security and privacy still remain top concern for businesses. /
云安全和隐私依然是业务中考虑最多的要素。/
A majority of respondents (56%) say their number one challenge for migrating / to the cloud /
多数应答者（56%）提到他们最大的忧虑 / 迁移到云 /
is an uncertain ability / to enforce their security policies / at cloud provider sites [8]. /
是一种不确定的能力 / 将实施安全政策的 / 在云供应商网站。/

警惕云计算风险

Beware the Risks of Cloud Computing

Despite many benefits, cloud computing has its own risks. Companies need to be fully aware of these before adopting the technology.

Cloud Outage

Companies store information and data in the cloud so they can be accessed anytime from anywhere using any devices. However, there are times when the system fails or underperforms. As with any information technology, cloud-based services are prone to outages and service interruptions. Even the best cloud service providers have faced these issues in the past, despite having high standards of maintenance. In 2014 alone, Google, Samsung, Adobe, Amazon, Microsoft, Yahoo, VMware, DropBox, Tumblr, and Facebook all experienced cloud outages ranging from a few hours to a few days [1].

Security

Security is one of the biggest concerns of cloud computing because the cloud is essentially the Internet. Your cloud server is accessible over the public Internet, which means your company's sensitive information is accessible as well. It's a potential goldmine for cybercriminals.

Privacy

Privacy is another concern that prevents companies from adopting cloud services. Your cloud service providers have full access to your personal data, which might be shared with third parties without your consent. In fact, many cloud providers can share information with third parties if necessary for purposes of law and order even without a warrant. That's often stated in their privacy policies, which users have to agree to before using the services.

Hidden Costs

Cloud advocates argue it makes good financial sense to move to the cloud [2]. However, a survey of 468 CIOs revealed that 78% of them worry about potential hidden costs. Here are the top concerns of CIOs [3]:

- Poor end user experience due to performance bottlenecks (64 percent).
- The impact of poor performance on brand perception and customer loyalty (51 percent).
- Loss of revenue due to poor availability, performance, or troubleshooting (44 percent).
- Increased costs of resolving problems in a more complex environment (35 percent).
- Increased effort required to manage vendors and service level agreements (23 percent).

Conclusion

Every technology has its pros and cons. A particular technology might be a great asset to your company when used properly, but it might also cause damage if misunderstood or misused. Evaluate the technology carefully before jumping on the bandwagon.

出 处

1. The worst cloud outages of 2014 (so far), http://goo.gl/JQG6WC
2. 5 Financial Benefits of Moving to the Cloud, http://goo.gl/J7qiiM
3. Hidden Costs Of Cloud Computing, Revealed, http://goo.gl/AQ8zy

核心语法

熟悉 # 关系代名词。# 关系代名词连接两个句子，兼具 # 连词和 # 代名词的作用。例文将介绍多种关系代名词。# 关系代名词有章可循，理解原理并参照例句学习将事半功倍。

- Privacy is another **concern** (**that** prevents companies from adopting cloud services). 隐私是另一大忧虑/它阻止公司选择云服务。
- That's stated in **their privacy policies**, **which** users have to agree to. 这被标注在它们的个人隐私政策内，用户必须同意。
- **Your cloud server is accessible over the public Internet**, **which** means your company's sensitive information is accessible as well. 用公共网络可以登录你的云服务器，这意味着/可以获取你公司的敏感信息/也。
- All companies experienced **cloud outages** ([**which** were] ranging from a few hours to a few days). 所有公司都经历过云中断/从几小时到几天不等。

熟悉 # 分词从句。# 分词从句是缩短句子的一种方式。理解 # 分词从句需先找出分词的主语。例文将介绍最常用的 # 分词从句。过去分词前的 being 可省略。

- Companies need to be aware of these **before adopting** the technology. 公司需要意识到这些/在采纳技术之前。

  = Companies need to be aware of these **before they adopt** the technology. =公司需要意识到这些/在它们（=公司）采纳技术之前。
- A particular technology might be a great asset **when** [**being**] **used** properly. 独特的技术可能是一笔巨大的资产/被合理使用时。

  = A particular technology might be a great asset **when it is used** properly. =独特的技术可能是一笔巨大的资产/它（=特定技术）被合理使用时。

熟悉 # 前置词，使用方法类似 # 连词，但 # 前置词后不接句子。

- **Despite** many benefits, cloud computing has its own risks. 尽管有很多益处，但云计算自身存在风险。
- **As with** any information technology, cloud–based services are prone to outages. 和其他信息技术一样，云服务也易中断。
- Even the best service providers have faced these issues in the past, **despite** having high standards of maintenance. 甚至是最优秀的服务供应商也面临过这些问题/在过去，尽管有高水平的维修。
- COIs worry about loss of revenue **due to** poor performance. 首席信息官担心/损失利润/较差的性能引起的。

单词&短语

beware 小心，注意
despite ~ 不管 ~
benefit 利益，长处
risk 风险
be aware of ~ 意识到 ~
adopt 采纳，选择
outage 中断
be prone to ~ 易于 ~
interruption 妨碍，中断
face 面对
in the past 在过去
standard 标准
maintenance 维修，管理
experience 经历
range from A to B（范围）从 A 到 B
concern 担心，忧虑
essentially 本质上
accessible 可访问的
sensitive 敏感的
as well 也
potential 潜在的
goldmine 金矿
prevent A from ~ing 阻止 A 做 ~
access to ~ 访问 ~
personal 个人的，私人的
share A with B 和 B 共享 A
consent 同意
in fact 事实上
if necessary 如果需要
for purposes of ~ 目的为 ~
law and order 治安
warrant 授权证
state 提及，陈述
privacy policy 隐私保护政策
agree to ~ 同意 ~
hidden 隐藏的
advocate 提倡，鼓吹
argue 主张
make sense 有意义，讲得通
reveal 揭露
worry about ~ 担心 ~
potential 潜在的
use experience 用户体验
due to ~ 因为 ~
bottleneck 瓶颈
impact 影响
brand perception 品牌意识
customer loyalty 顾客忠诚度

根据提示完成句子

____
隐私

________ is another ________ /
隐私保护是另一大考虑/

that ________ companies ____ _________ cloud services. /
这阻止了公司选择云服务。/

Your cloud service providers have ____ ______ /
你的云服务供应商可以全权访问/

to your personal data, / which might be shared /
你的个人数据，/ 可能被共享/

with third parties / _______ ____ _______. /
与第三方/ 未经你的同意。/

__ ____, many cloud providers can share information /
实际上，众多云供应商可以共享信息/

with third parties / __ _________ for purposes of ___ ___ _____ /
与第三方/ 如果治安需要/

even _______ _ ________. /
甚至没有授权书。/

That's often ______ / in their _______ _________, /
这经常被标注在/ 它们的个人信息保护政策中，/

which users have to agree to / before using the services. /
用户必须同意/ 在使用服务之前。/

__________ /
结论

Every technology has its ____ ___ ____. /
所有技术都有其优缺点。/

A particular technology might be a great asset /
特殊技术可能是一项巨大资源/

to your company / when used properly, /
对你的公司而言/ 被合理使用时，/

but it might also _____ ______ / if _____________ or _______. /
但也可能导致损失/ 如果被误解或者误用。/

________ the technology carefully / ______ _______ __ _ _________. /
认真评估技术/ 在追赶潮流之前。/

思考题

解答题

1 （理解）以下哪一项是云特有的风险？

ⓐ 安全 ⓑ 侵犯隐私

ⓒ 网络服务障碍 ⓓ SLA

2 （论述）总结云的隐藏成本。

讨论

1 调查云服务缺陷案例，评价服务业的应对方法。

2 调查如何克服云的不足之处。

课堂总结

1 学习#关系代名词的概念和例句。

2 学习#分词从句的概念和例句。

3 学习#连词、#前置词、#接续副词。

答案

1 d.（亲自构建IT基础设施时，并没有SLA这个概念。） 2 性能瓶颈导致最终用户体验恶劣，影响品牌认知和顾客忠诚度的低劣性能带来的影响，可用性、性能、问题解决能力等不足导致销量减少，在复杂环境中解决问题的成本增加，公司和SLA需要更努力地进行管理。

 翻译

Beware the Risks / of Cloud Computing
警惕云计算风险

Despite many benefits, / cloud computing has its own risks. /
尽管有很多益处，/ 但云计算自身存在风险。/
Companies need to be fully aware of these /
公司需要充分意识到这些 /
before adopting the technology. /
在选择技术之前。/

Cloud Outage / 云中断

Companies store information and data / in the cloud / so they can be accessed / anytime /
公司保存信息和数据 / 到云 / 所以信息可以被访问 / 随时 /
from anywhere / using any devices. / However, / there are times /
随地 / 用任何设备。/ 但是，/ 有时 /
when the system fails / or underperforms. / As with any information technology, /
系统失灵 / 或表现不佳。/ 和其他信息技术一样，/
cloud-based services are prone to outages and service interruptions. /
云服务也易中断和受干扰。/
Even the best cloud service providers have faced these issues / in the past, /
甚至最优秀的服务供应商也面临过这些问题 / 在过去，/
despite having high standards of maintenance. /
尽管有高水平的维修。/
In 2014 alone, / Google, Samsung, Adobe, Amazon, Microsoft, Yahoo, VMware, DropBox, Tumblr,
仅 2014 年，/ 谷歌、三星、Adobe、亚马逊、微软、雅虎、VMware、DropBox、Tumblr
and Facebook all experienced cloud outages / ranging from a few hours to a few days [1]. /
和 Facebook 都经历过云中断 / 从几小时到几天不等。/

Security / 安全

Security is one of the biggest concerns / of cloud computing /
安全是最大的忧虑之一 / 云计算的 /
because the cloud is essentially the Internet. /
因为云本质上是网络。/
Your cloud server is accessible / over the public Internet, / which means /
你的云服务器是可访问的 / 通过公共网络，/ 这意味着 /
your company's sensitive information is accessible /
你公司的敏感信息是可访问的 /
as well. / It's a potential goldmine / for cybercriminals. /
也。/ 这是一个潜在的金矿 / 对网络罪犯而言。/

Privacy / 隐私

Privacy is another concern / that prevents companies from adopting cloud services. /
隐私保护是另一大忧患 / 这阻止了公司选择云服务。/
Your cloud service providers have full access / to your personal data, / which might be shared /
你的云服务供应商可以全权访问 / 你的个人数据，/ 这可能被共享 /
with third parties / without your consent. / In fact, many cloud providers can share information /
与第三方 / 未经你的同意。/ 实际上，众多云供应商可以共享信息 /
with third parties / if necessary for purposes of law and order / even without a warrant. /
与第三方 / 如果治安需要 / 甚至没有授权书。/
That's often stated / in their privacy policies, / which users have to agree to / before using the services. /
这经常被标注在 / 它们的个人信息保护政策中，/ 用户必须同意 / 在使用服务之前。/

Hidden Costs / 隐藏成本

Cloud advocates argue / it makes good financial sense / to move to the cloud [2]. /
云支持者认为 / 从财政角度讲得通 / 迁移到云。/
However, a survey of 468 CIOs revealed / that 78% of them worry about potential hidden costs. /
但 468 位首席信息官调查结果显示 / 其中 78% 的人担心隐藏成本。/
Here are the top concerns of CIOs [3]: /
以下为首席信息官的最大忧虑：/

- Poor end user experience / due to performance bottlenecks (64 percent). /
  不良的用户体验/因性能瓶颈导致的（64%）。/
- The impact of poor performance / on brand perception and customer loyalty (51 percent). /
  不良的性能影响/对品牌认知和顾客忠诚度造成的（51%）。/
- Loss of revenue / due to poor availability, performance, or troubleshooting (44 percent). /
  利润减少/因可用性、性能、问题解决能力等不足（44%）。/
- Increased costs of resolving problems / in a more complex environment (35 percent). /
  解决问题成本增加/在更加复杂的环境中（35%）。/
- Increased effort required / to manage vendors and service level agreements (23 percent). /
  需要更多努力/为管理厂商和SLA（23%）。/

Conclusion / 结论

Every technology has its pros and cons. / A particular technology might be a great asset /
所有技术都有其优缺点。/ 特殊技术可能是一项巨大的资源 /
to your company / when used properly, / but it might also cause damage /
对你的公司而言 / 被合理使用时，/ 但这也可能导致损失 /
if misunderstood or misused. / Evaluate the technology carefully /
如果被误解或者误用。/ 认真评估技术 /
before jumping on the bandwagon. /
在追赶潮流之前。/

常用云计算术语集锦

Essential Cloud Computing Terms

New technology comes with new terminology, and it can be confusing. Here are some common cloud computing terms that will help you follow the industry trends and development activities [1, 2].

Amazon EC2

Amazon's IaaS offering. Short for Amazon Elastic Compute Cloud. See *IaaS*.

Amazon S3

Amazon's cloud storage service. Short for Amazon Simple Storage Services.

Amazon Web Services (AWS)

A collection of remote computing services that make up Amazon's cloud computing platform. The well-known services are *Amazon EC2* and *Amazon S3*.

Cloud App

An application that runs on a cloud platform and is accessed via the Internet. Short for *cloud application*.

Cloud computing

The delivery of computing as a service rather than a product.

Cloud portability

The ability to move applications or services from one cloud provider to another.

Cloudsourcing

Replacing traditional IT services with cloud services.

Elastic computing

The process of provisioning and deprovisioning resources in an autonomic manner in order to adapt to workload changes.

Google Compute Engine (GCE)
Google's IaaS offering. See *IaaS*.

Hybrid cloud
A cloud computing model that combines two or more cloud computing models. See *public cloud* and *private cloud*.

Infrastructure as a service (IaaS)
A cloud computing model through which computing infrastructure is delivered over the Internet. The infrastructure includes virtualized servers, storage, hardware, and software.

Microsoft Azure
Microsoft's PaaS and IaaS offering. See *PaaS* and *IaaS*.

Platform as a service (PaaS)
A cloud computing model through which a computing platform is delivered as a service to users. The model allows developers to build software solutions over the Internet. The platform includes operating system, development environment, database, and web server.

Private cloud
A cloud computing model in which services and infrastructure are operated solely for a single organization. They can be managed internally or by a third party, and hosted either internally or externally.

Public cloud
A cloud computing model in which services are open for public use over the Internet.

Software as a service (SaaS)
A cloud computing model that provides access to application software and databases over the Internet. Cloud providers manage the infrastructure and platforms that run the applications. SaaS is sometimes referred to as "on-demand software".

出 处

1. Cloud Dictionary: 50 Cloud Computing Terms to Know, http://goo.gl/PYCai1
2. Cloud Computing Glossary, http://goo.gl/BA0wal

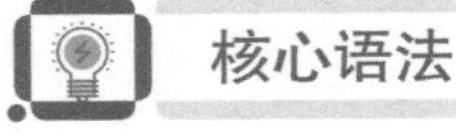

核心语法

熟悉 # 关系代名词。# 关系代名词连接两个句子，兼具 # 连词和 # 代名词的作用。# 关系代名词有章可循，理解原理并参照例句学习将事半功倍。

- Here are some common **terms** (**which** will help you follow the industry trends). 以下为一些常用术语/有助于你紧跟业界趋势。
- Cloud App is **an application** (**that** runs on a cloud platform). Cloud APP是一款应用程序/在云平台运行的。

熟悉 # 前置词 + 关系代名词。因 # 关系代名词前有 # 前置词，故较难理解。

- IaaS is **a cloud computing model**. + Computing infrastructure is delivered **through the model**. Iaas是一种云计算模型 + 计算基础设施被传输/通过这个模型。

  = IaaS is **a cloud computing model** (**which** computing infrastructure is delivered **through**). =Iaas是一种云计算模型/通过它计算基础设施被传输。

  = IaaS is **a cloud computing model** (**through which** computing infrastructure is delivered).
- Private Cloud is **a cloud computing model**. + Services are operated for a single organization **in the model**. 私有云是一种云计算模型 + 服务被运营为了一个单独组织/在模型中。

  = Private Cloud is **a cloud computing model** (**which** services are operated for a single organization **in**). =私有云是一种云计算模型/在其中服务为了一个单独组织运营。

  = Private Cloud is **a cloud computing model** (**in which** services are operated for a single organization).

单词&短语

essential 必需的
term 术语
terminology 术语
confusing 令人困惑的
follow 跟随
trend 趋势
activity 活动
offering 提供的东西
short for ~ 是 ~ 的简称
remote 远程
make up 构成
well-known 众所周知的
access 访问
delivery 传送
rather than 而不是
portability 可移植性
replace 替换
traditional 传统的
elastic 有弹力的
provisioning 赋予权限
deprovisioning 取消权限
in an autonomic manner 自主地
in order to V 为了 ~
adapt to ~ 适应 ~
combine 结合
include 包括
allow O to V 允许 O 做 V
operating system 操作系统
development environment 开发环境
database 数据库
web server Web 服务器
operate 运营
solely 仅仅，唯一地
organization 组织
internally 内部地
third party 第三方
externally 外部地
open for ~ 为 ~ 开放
provide 提高
Refer A as B 将 A 称为 B
on-demand 按需

根据提示完成句子

___ ____________ ______ _____ ___ _____________, /
新技术伴随着新术语而来，/

and it can __ ____________. /
可能让人感到困惑。/

Here are some common cloud computing terms /
以下为一些常用的云计算术语/

that will help you _____ the industry ______ and __________ __________. /
将有助于你紧跟业界趋势和研发活动。/

Platform as a service (PaaS) /
平台即服务（PaaS）

A cloud computing model /
一种云计算模式/

________ _____ a computing platform is delivered /
通过它，云计算平台被传输/

__ _ _______ / to users. / The model allows developers /
作为服务/ 给用户。/ 这个模型让研发人员/

to build software solutions / ____ ___ ________. /
构建软件解决方案/ 在网上。/

The platform includes
平台包括

_________ ______, __________ ___________, database, and web server. /
操作系统、研发环境、数据库、Web服务器。/

思考题

解答题

1 （理解）以下哪项术语与其他三项不同？

ⓐ 亚马逊 S3 ⓑ 微软 Azure

ⓒ 谷歌 Compute Engine ⓓ 亚马逊 EC2

2 （论述）简述众包和外包的不同。

讨论

1 亲自制作云相关字典（提示：http://goo.gl/QESAAQ）。

课堂总结

1 学习#关系代名词的概念和例句。

2 学习#前置词+关系代名词的概念和例句。

3 学习#关系副词的概念和例句。

答案

1 a.（亚马逊S3为存储服务，其他具有IaaS特征。） 2 外包指雇佣人员提供服务，众包指一家公司或机构将过去由员工执行的工作任务，以自由、自愿的形式外包给非特定的（而且通常是大型的）大众网络的做法。

翻译

Essential Cloud Computing Terms
常用云计算术语集锦

New technology comes with new terminology, / and it can be confusing. /
新技术伴随着新术语而来 / 这可能让人感到很困惑。/
Here are some common cloud computing terms /
以下为一些常用的云计算术语 /
that will help you follow the industry trends and development activities [1, 2]. /
将有助于你紧跟业界趋势和研发活动。/

Amazon EC2 / 亚马逊EC2

Amazon's IaaS offering. / Short for Amazon Elastic Compute Cloud. / See IaaS.
亚马逊的 IaaS。/ 是 Amazon Elastic Compute Cloud 的缩写。/ 参见 IaaS。/

Amazon S3 / 亚马逊S3

Amazon's cloud storage service. / Short for Amazon Simple Storage Services. /
亚马逊的云存储服务。/ 是 Amazon Simple Storage Services 的缩写。/

Amazon Web Services (AWS) / 亚马逊Web服务（AWS）

A collection of remote computing services / that make up Amazon's cloud computing platform. /
远程计算服务的集合 / 组成了亚马逊的云计算平台。/
The well-known services are Amazon EC2 and Amazon S3. /
众所周知的服务是亚马逊 EC2 和亚马逊 S3。/

Cloud App / Cloud App

An application / that runs on a cloud platform / and is accessed / via the Internet. /
一款应用 / 在云平台运行的。/ 被访问 / 通过网络。/
Short for cloud application. /
是 Cloud Application 的缩写。/

Cloud computing / 云计算

The delivery of computing as a service / rather than a product. /
以服务形式提供计算 / 而不是产品。/

Cloud portability / 云移植性

The ability to move applications or services / from one cloud provider to another. /
移动应用或者服务的能力 / 从一个云供应商到另一个。/

Cloudsourcing / 众包

Replacing traditional IT services with cloud services. /
用云服务取代传统 IT 服务。/

Elastic computing / 弹性计算

The process of provisioning and deprovisioning resources / in an autonomic manner /
对资源访问授予权限和取消权限的过程 / 自主地 /

in order to adapt to workload changes. /
为适应工作负载变化。/

Google Compute Engine (GCE) / 谷歌Compute Engine（GCE）

Google's IaaS offering. / See IaaS. /
谷歌的 IaaS。/ 参见 IaaS。/

Hybrid cloud / 混合云

A cloud computing model / that combines two or more cloud computing models. /
一种云计算模型 / 结合了两种或两种以上云计算模型。/
See public cloud and private cloud. /
参见“公共云”和“私有云”。/

Infrastructure as a service (IaaS) / 基础设施即服务（IaaS）

A cloud computing model / through which computing infrastructure is delivered /
一种云计算模型 / 通过其传输计算基础设施 /
over the Internet. /The infrastructure includes virtualized servers, storage, hardware, and software. /
在网上。/ 基础设施包括虚拟服务器、内存、硬件、软件。/

Microsoft Azure / 微软Azure

Microsoft's PaaS and IaaS offering. / See PaaS and IaaS. /
微软的 Paas 和 Iaas。/ 参见 Paas 和 IaaS。/

Platform as a service (PaaS) / 平台即服务（PaaS）

A cloud computing model / through which a computing platform is delivered /
一种云计算模型 / 通过其传输云计算平台 / 作为服务 / 给用户。/
as a service / to users. / The model allows developers / to build software solutions / over the Internet. /
这个模型让研发人员 / 构建软件解决方案 / 在网上。/
The platform includes operating system, development environment, database, and web server. /
平台包括操作系统、研发环境、数据库、Web 服务器。/

Private cloud / 私有云

A cloud computing model / in which services and infrastructure are operated / solely for a single organization. /
一种云计算模型 / 在这里运营服务和基础设施 / 仅仅为一个组织。/
They can be managed / internally / or by a third party, / and hosted / either internally / or externally. /
它们可以被管理 / 在内部 / 或者通过第三方 / 并且被托管 / 在内部 / 或者外部。/

Public cloud / 公共云

A cloud computing model / in which services are open / for public use / over the Internet. /
一种云计算模型 / 在此开放服务 / 给大众使用 / 通过网络。/

Software as a service (SaaS) / 软件即服务（SaaS）

A cloud computing model / that provides access / to application software and databases /
一种云计算模型 / 可访问 / 应用软件和数据库 /
over the Internet. / Cloud providers manage the infrastructure and platforms /
在网上。/ 云供应商管理基础设施和平台 /
that run the applications. / SaaS is sometimes referred to / as “on-demand software”.
运行应用的。/SaaS 有时称为 / “按需软件”。/

经典语录之云计算篇

Notable Quotes on Cloud Computing

Here are some notable quotes on cloud computing.

"First to mind when asked what 'the cloud' is, a majority respond it's either an actual cloud, the sky, or something related to weather."

— **Citrix Cloud Survey Guide** (August 2012)

"I don't need a hard disk in my computer if I can get to the server faster... carrying around these non-connected computers is byzantine by comparison."

— **Steve Jobs**, late chairman of Apple

"Cloud is about how you do computing, not where you do computing."

— **Paul Maritz**, VMware CEO

"Every kid coming out of Harvard, every kid coming out of school now thinks he can be the next Mark Zuckerberg, and with these new technologies like cloud computing, he actually has a shot."

— **Marc Andreessen**, Board Member of Facebook

"We believe we're moving out of the Ice Age, the Iron Age, the Industrial Age, the Information Age, to the participation age. You get on the Net and you do stuff. You IM (instant message), you blog, you take pictures, you publish, you podcast, you transact, you distance learn, you telemedicine. You are participating on the Internet, not just viewing stuff. We build the infrastructure that goes in the data center that facilitates the participation age. We build that big friggin' Webtone switch. It has security, directory, identity, privacy, storage, compute, the whole Web services stack."

— **Scott McNealy**, former CEO, Sun Microsystems

"Cloud computing is often far more secure than traditional computing, because companies like Google and Amazon can attract and retain cyber-security personnel of a higher quality than many governmental agencies."

— **Vivek Kundra**, former federal CIO of the United States

"There was a time when every household, town, farm or village had its own water well. Today, shared public utilities give us access to clean water by simply turning on the tap; cloud computing works in a similar fashion. Just like water from the tap in your kitchen, cloud computing services can be turned on or off quickly as needed. Like at the water company, there is a team of dedicated professionals making sure the service provided is safe, secure and available on a 24/7 basis. When the tap isn't on, not only are you saving water, but you aren't paying for resources you don't currently need."

— **Vivek Kundra**, former federal CIO of the United States.

"The interesting thing about cloud computing is that we've redefined cloud computing to include everything that we already do. I can't think of anything that isn't cloud computing with all of these announcements. The computer industry is the only industry that is more fashion-driven than women's fashion. Maybe I'm an idiot, but I have no idea what anyone is talking about. What is it? It's complete gibberish. It's insane. When is this idiocy going to stop?" — **Larry Ellison**, chairman, Oracle

"If you think you've seen this movie before, you are right. Cloud computing is based on the time-sharing model we leveraged years ago before we could afford our own computers. The idea is to share computing power among many companies and people, thereby reducing the cost of that computing power to those who leverage it. The value of time share and the core value of cloud computing are pretty much the same, only the resources these days are much better and more cost effective." — **David Linthicum**,
author of *Cloud Computing and SOA Convergence in Your Enterprise: A Step-by-Step Guide*

核心语法

#省略是使句子复杂的原因之一。#省略包括符合语法规则的省略和与语法不相关的省略。牢牢掌握前者后，“句读百遍，后者义自见”。

- [What comes] First to mind when [people are] asked what ‘the cloud’ is, a majority [of people] respond [that] it’s either an actual cloud, the sky, or something [which is] related to weather. 最先想起/当人们被问及/何为“云”/多数人回应/是实际的云彩、天空/或一些/与天气相关的。
- Every kid [who is] coming out of Harvard, every kid [who is] coming out of school now thinks [that] he can be the next Mark Zuckerberg, and with these new technologies like cloud computing, he actually has a shot. 所有孩子/哈佛毕业的/所有孩子/毕业的/现在认为/自己会成为下一个马克・扎克伯格/并且用这些新技术/诸如云计算/实际上有可能。
- There is a team of dedicated professionals [who are] making sure [that] the service [which is] provided is safe. 有专家组/确定/被提供的服务/是安全的。

熟悉#that用法。例文将介绍使用that的多种从句。无需死记硬背语法，理解句子原理即可。以下例句提及了语法术语，帮助大家轻松找到类似例句。

- We build **the infrastructure that** goes in **the data center that** facilitates the participation age. 我们创建基础设施/进入数据中心的/促进参与时代的。
  - ▶ 熟悉**#关系代名词that**。that从句中还有that从句。
- The interesting thing about cloud computing is **that** we’ve redefined cloud computing to include **everything that** we already do. 关于云计算有趣的是/我们已经重新定义了云计算/以包括所有事情/我们已经做过的。
  - ▶ 熟悉**#连词that**和**#关系代名词that**。that从句中还有that从句。

- The idea is to share computing power among many companies and people, thereby reducing the cost of **that** computing power. 这个想法是/分享计算能力/与众多公司和个人，/所以它可以降低计算能力的成本。
  ▶ 熟悉#指示形容词that。指代前文名词。

单词&短语

notable 值得关注的
come to mind 想起
majority 多数
either A or B 或者 A，或者 B
related to ~ 与 ~ 相关的
byzantine 拜占庭帝国的
by comparison 相比之下
have a shot 有可能
Ice Age 冰川时期
Iron Age 铁器时代
Industrial Age 工业时代
Information Age 信息时代
participation 参与
do stuff 显身手，做分内的事
transact 交易
distant learn 远程教育
telemedicine 远程医疗
facilitate 促进
frigging 受诅咒的
identity 身份
household 家政
village 村落
water well 井
public utility（水、电、煤气等）公共事业
turn on(off) a tap 打开 / 关闭阀门
in a(n) ~ fashion 以 ~ 方式
as needed 根据需要
dedicated 专注于
make sure 确认
provide 提供
not only ~ but (also) ... 不仅……而且……
pay for 支付
currently 现在
redefine 重新定义
include 包括
announcement 发布，公布
-driven ~ 主导的
idiot 傻瓜
complete 完全的
gibberish 胡扯
insane 愚蠢的，疯狂的
idiocy 愚蠢
based on 以 ~ 为基础
leverage 使用
afford 担负得起
thereby 因此
value 价值
core value 核心价值
pretty much 相当，几乎
cost effective 划算的

课堂总结

[1] 学习#关系代名词、#关系副词的单词和例句。

[2] 学习#分词从句的概念和例句。

[3] 学习#倒装和#省略。

根据提示完成句子

"_____ ___ _ _____ /

"有一段时期/

when every household, town, farm or village had its own water well. /

当所有家庭、城镇、农场或者村落都有它们自己的水井。/

Today, ______ _______ ________ give us access / to clean water /

今天，共享公共事业让我们可以接近/ 更干净的水/

by simply turning on the tap; /

只需要打开水龙头；/

cloud computing works / in a _______ _______. /

云计算运转/ 以同样的方式/

Just like water from the tap / in your kitchen, /

就像水龙头中流出的水/ 在你的厨房，/

cloud computing services can be ______ __ __ ___ / quickly /

云计算服务可以开关/ 迅速地/

__ _______. / Like at the water company, /

根据需要。/ 就像自来水公司，/

there is a team of __________ professionals / ______ ____ /

有专任专家组/ 保证/

the service provided is safe, secure and available / on a __ /_ basis. /

被提供的服务是安全的、可靠的、可利用的/ 按照24小时7天的标准。/

When the tap isn't on, / ___ ___ are you ______ _____, /

如果不关水龙头，/ 你不仅在浪费水，/

___ you aren't ______ __ resources you don't currently need." /

而且还没有为你不需要的资源付款。" /

— **Vivek Kundra**, former federal CIO of the United States.

翻译

Notable Quotes / on Cloud Computing
经典语录之云计算篇

Here are some notable quotes / on cloud computing. /
以下是一些经典语录 / 关于云计算的。/

"First to mind / when asked / what 'the cloud' is, / a majority respond /
"最先想到的是 / 当被问及 / 何为"云"时 / 大多数人回答 /
it's either an actual cloud, the sky, or something / related to weather." /
实际的云彩、天空或者一些 / 与天气有关的。" /

——Citrix 云调查指南（2012 年 8 月）

"I don't need a hard disk / in my computer / if I can get to the server faster… /
"我不需要硬盘 / 在我的电脑里 / 如果我能更迅速地登录服务器……/
carrying around these non-connected computers is byzantine / by comparison." /
携带这些不联网的电脑四处走动仿佛回到了拜占庭帝国时期 / 通过对比。" /

——史蒂夫·乔布斯，前苹果公司行政总裁

"Cloud is about how you do computing, / not where you do computing." /
"云是关于你如何计算的，/ 而不是在哪里计算。" /

——保罗·马瑞兹，VMware 首席执行官

"Every kid coming out of Harvard, / every kid coming out of school /
"所有孩子 / 毕业于哈佛的 / 所有孩子 / 毕业的 /
now thinks / he can be the next Mark Zuckerberg, / and with these new technologies /
现在认为 / 自己将是下一个马克·扎克伯格 / 用这些新技术 /
like cloud computing, / he actually has a shot." /
诸如云计算 / 实际上有可能。" /

——马克·安德森，Facebook 董事会成员

"We believe / we're moving out of the Ice Age, the Iron Age, the Industrial Age,
"我们相信 / 我们将走出冰川时期、铁器时代、工业时代、
the Information Age, to the participation age. /
信息时代，走向参与时代。/
You get on the Net and you do stuff. / You IM (instant message), / you blog, / you take pictures, /
你上网大显身手。/ 使用即时通信工具、/ 登录博客、/ 拍照、/

you publish, / you podcast, / you transact, / you distance learn, / you telemedicine. /
出书、/ 登录播客、/ 交易、/ 接受远程教育 / 和远程医疗。/
You are participating on the Internet, / not just viewing stuff. / We build the infrastructure /
你是互联网的参与者，/ 并不仅是袖手旁观。/ 我们构建基础设施 /
that goes in the data center / that facilitates the participation age. /
进入数据中心的 / 促进参与时代的。/
We build that big friggin' Webtone switch. /
我们构建了那个大得该死的 Webtone 转换器。/
It has security, directory, identity, privacy, storage, compute, the whole Web services stack." /
它具有安全、目录、身份、隐私、存储器、计算、Web 服务全栈。"/

——**史考特·麦克里尼**，前 Sun Microsystems 首席执行官

"Cloud computing is often far more secure / than traditional computing, /
"云计算通常更安全 / 比传统计算，/
because companies like Google and Amazon can attract and retain cyber-security personnel /
因为谷歌和亚马逊这类公司可以吸引和留住安全人才 /
of a higher quality / than many governmental agencies."
更优质的 / 相比许多政府机构。"/

——**维维克·昆德拉**，前美联邦政府首席信息官

"There was a time / when every household, town, farm or village had its own water well. /
"有一段时期 / 所有家庭、城镇、农场或者村落都有自己的水井。/
Today, shared public utilities give us access / to clean water / by simply turning on the tap; /
今天，共享公共事业让我们可以接近 / 更干净的水 / 只需要打开水龙头；/
cloud computing works / in a similar fashion. / Just like water from the tap / in your kitchen, /
云计算运行 / 以同样的方式 / 就像水龙头中流出的水 / 在你的厨房，/
cloud computing services can be turned on or off / quickly / as needed. / Like at the water company, /
云计算服务可以开关 / 迅速地 / 根据需要。/ 就像自来水公司，/
there is a team of dedicated professionals / making sure /
有专任专家组 / 保证 /
the service provided is safe, secure and available /
被提供的服务是安全的、可靠的、可利用的 /
on a 24/7 basis. / When the tap isn't on, / not only are you saving water, /
按照 24 小时 7 天的标准。/ 如果不关水龙头，/ 你不仅在浪费水，/
but you aren't paying for resources you don't currently need." /
而且没有为你不需要的资源付款。"/

——**维维克·昆德拉**，前美联邦政府首席信息官

"The interesting thing about cloud computing is / that we've redefined cloud computing /
"关于云计算有趣的事情是 / 我们已经重新定义云计算 /
to include everything / that we already do. / I can't think of anything / that isn't cloud computing /
以包含一切 / 我们已经做过的。/ 我无法想起任何事情 / 不是云计算的 /
with all of these announcements. /
在这所有成果中。/
The computer industry is the only industry / that is more fashion-driven /
计算机领域是唯一的领域。/ 更加推动潮流的 /
than women's fashion. / Maybe I'm an idiot, / but I have no idea / what anyone is talking about. /
比女性时尚。/ 可能我是傻子，/ 但我不知道 / 别人在谈论什么。/
What is it? / It's complete gibberish. / It's insane. / When is this idiocy going to stop?" /
那是什么呢？ / 这完全是胡扯。/ 这非常疯狂 / 这种愚蠢的行为何时才能停止呢?" //

——**拉里·埃里森**，甲骨文公司总裁

"If you think / you've seen this movie before, / you are right. /
"如果你认为 / 以前看过这部电影，/ 那你是对的。/
Cloud computing is based on the time-sharing model /
云计算是基于分时的模式 /
we leveraged years ago / before we could afford our own computers. /
很多年前使用的 / 在我们可以使用自己的计算机之前。/
The idea is / to share computing power /
这个想法就是 / 分享计算能力 /
among many companies and people, / thereby reducing the cost of that computing power /
在众多公司和人群中，/ 所以减少计算能力的成本 /
to those who leverage it. / The value of time share / and the core value of cloud computing /
给那些使用的人们。/ 分时的价值 / 和云计算的核心价值 /
are pretty much the same, / only the resources these days are much better and more cost effective." /
相当类似，/ 只是现在的资源更加优质和划算。"/

——**戴维·林西克姆**，*Cloud Computing and SOA Convergence in Your Enterprise: A Step-by-Step Guide* 作者

第六部分

实战

本部分以前文内容为基础，认真学习后足以攻破IT技术文章阅读。

维基百科上的相关叙述
According to Wikipedia

Here's how Wikipedia defines each tech topic we have discussed in this book.

Internet Security

Internet security is a tree branch of computer security specifically related to the Internet, often involving browser security but also network security on a more general level as it applies to other applications or operating systems on the whole. Its objective is to establish rules and measures to use against attacks over the Internet. The Internet represents an insecure channel for exchanging information leading to a high risk of intrusion or fraud, such as phishing. Different methods have been used to protect the transfer of data, including encryption [4].

Robotics

Robotics is the branch of mechanical engineering, electrical engineering and computer science that deals with the design, construction, operation, and application of robots, as well as computer systems for their control, sensory feedback, and information processing [6].

Artificial Intelligence

Artificial intelligence (AI) is the intelligence exhibited by machines or software. It is an academic field of study which studies the goal of creating intelligence. Major AI researchers and textbooks define this field as "the study and design of intelligent agents", [1] where an intelligent agent is a system that perceives its environment and takes actions that maximize its chances of success. John McCarthy, who coined the term in 1955, defines it as "the science and engineering of making intelligent machines."[5]

Big Data

Big data is a broad term for data sets so large or complex that traditional data processing applications are inadequate. Challenges include analysis, capture, curation, search, sharing, storage, transfer, visualization, and information privacy. The term often refers simply to the use of predictive analytics or other certain advanced methods to extract value from data, and seldom to a particular size of data set [2].

Internet of Things

The Internet of Things (IoT) is the network of physical objects or "things" embedded with electronics, software, sensors and connectivity to enable it to achieve greater value and service by exchanging data with the manufacturer, operator and/or other connected devices. Each thing is uniquely identifiable through its embedded computing system but is able to interoperate within the existing Internet infrastructure [3].

Cloud Computing

Cloud computing is a computing term or metaphor that evolved in the late 2000s, based on utility and consumption of computing resources. Cloud computing involves deploying groups of remote servers and software networks that allow centralized data storage and online access to computer services or resources. Clouds can be classified as public, private or hybrid [1].

出处

1. Cloud computing, Wikipedia, http://goo.gl/3qxl
2. Big data, Wikipedia, http://goo.gl/DFFbr
3. Internet of Things, Wikipedia, http://goo.gl/pcrf
4. Internet security, Wikipedia, http://goo.gl/lEclW
5. Artificial intelligence, Wikipedia, http://goo.gl/pcGe
6. Robotics, Wikipedia, http://goo.gl/VRL2h

单词&短语

define 定义
discuss 讨论，议论
term 术语
metaphor 隐喻，比喻
evolve 进化
based on 基于～
utility 实用性
consumption 消耗
resource 资源
involve 包括
deploy 部署，安装
remote 远程
centralized 使集中的
be classified as 被归类为～
broad 广泛的
complex 复杂的
so ~ that ... 如此～以至于……
traditional 传统的
inadequate 不充分的
include 包括
analysis 分析
capture 捕捉
curation 综合处理
transfer 传输
visualization 可视化
refer to 提及
predictive analytics 预测分析（学）
advanced 先进的，高级的
method 方法
extract 提取
seldom 罕见，稀少
particular 特别的
physical object 物体
embedded 内置的
connectivity 连接（性）
enable A to V 使 A 能做 V
value 价值
exchange 交换
manufacture 制造商
operator 运营者
uniquely 独特地，唯一地
identifiable 可识别的
be able to V 能够做～
interoperate 联动
existing 既存的
branch 分支
specifically 特别地
related to 与～相关
involve 包括
general 一般的
apply to ～ 适用于～
objective 目的
establish 创建
measure 方案，方法，措施
attack 攻击
represent 代表
insecure 不安全的
exchange 交换
lead to ～ 导致～
intrusion 侵入
fraud 欺诈
phishing 网络欺诈
encryption 加密
exhibit 展示
academic 学术的
researcher 研究员
agent 代理商，中间商，中介
perceive 发觉
environment 环境
take action 采取措施
maximize 最大化
coin 创造（单词、术语）
mechanical engineering 机械工程
electrical engineering 电子工程

根据提示完成句子

__________ ____________ (AI) is the intelligence /

人工智能（AI）是智能/

_________ by machines or software. /

通过机器或者软件显示的。/

It is an ________ field of study /

它是学术领域的研究/

which studies the ____ of ________ ____________. /

研究创造智能的目的。/

Major AI ___________ and _________ define this field /

主要AI研究人员和教科书定义这个领域为/

as “the study and design of ___________ ______”, [1] /

“智能代理的研究与设计”，/

where an ___________ _____ is a system /

此处的“智能代理”是一个系统/

that _________ its ___________ and _____ _______ /

它感知环境并采取行动/

that ________ its chances of success. /

将成功的可能性最大化。/

John McCarthy, / who ______ the term / in 1955, / defines it as /

约翰·麦卡锡，/ 创造了这个术语/ 在1955年，/ 将其定义为/

“the _______ and ___________ of making intelligent machines.”[5] /

“创造智能机器的科学和工程”。/

翻译

According to Wikipedia
维基百科上的相关叙述

Here's how Wikipedia defines each tech topic / we have discussed in this book. /
以下是维基百科对各技术主题的定义 / 我们在本书中讨论过的。/

Internet Security / 互联网安全

Internet security is a tree branch of computer security / specifically related to the Internet, /
互联网安全是计算机安全的分支 / 特别与网络有关的，/
often involving browser security / but also network security / on a more general level /
通常包括浏览器安全 / 和网络安全 / 在更常规的水平上 /
as it applies to other applications or operating systems / on the whole. /
因为它适用于其他应用或操作系统 / 整体上。/
Its objective is to establish rules and measures / to use against attacks / over the Internet. /
其目的是制定规则和方法 / 以应对攻击 / 互联网上的。/
The Internet represents an insecure channel / for exchanging information /
网络代表不安全的渠道 / 为交流信息 /
leading to a high risk of intrusion or fraud, / such as phishing. /
导致高风险的入侵和欺诈 / 例如钓鱼网站。/
Different methods have been used / to protect the transfer of data, / including encryption [4]. /
不同的方法被用 / 以保护数据传输，/ 包括加密。/

Robotics / 机器人

Robotics is the branch of mechanical engineering, electrical engineering and computer science /
机器人是机器工程、电子工程计算机科学的分支 /
that deals with the design, construction, operation, and application of robots, /
它处理设计、结构、运营、机器人应用，/
as well as computer systems / for their control, sensory feedback, and information processing [6]. /
以及计算机系统 / 控制、传感器反馈、信息处理。/

Artificial Intelligence / 人工智能

Artificial intelligence (AI) is the intelligence / exhibited by machines or software. /
人工智能（AI）是智能 / 通过机器或者软件显示。/
It is an academic field of study / which studies the goal of creating intelligence. /
它是学术领域的研究 / 研究创造智能的目的。/
Major AI researchers and textbooks define this field / as "the study and design of intelligent agents", [1] /
主要 AI 研究人员和教科书定义这个领域为 / “智能代理的研究与设计”，/
where an intelligent agent is a system / that perceives its environment and takes actions /
此处的“智能代理”是一个系统 / 它感知环境并采取行动 /
that maximize its chances of success. / John McCarthy, / who coined the term / in 1955, /

将成功的可能性最大化。/ 约翰 · 麦卡锡 / 创造了这个术语 / 在 1955 年，/
defines it as / “the science and engineering of making intelligent machines.”[5] /
将其定义为 /“创造智能机器的科学和工程”。/

Big Data / 大数据

Big data is a broad term / for data sets / so large or complex /
大数据是宏观术语 / 数据集合 / 大且复杂 /
that traditional data processing applications are inadequate. /
以至于不再适合传统的数据处理应用。/
Challenges include / analysis, capture, curation, search, sharing, storage, transfer, visualization,
挑战包括 / 分析、捕捉、综合处理、搜索、分享、存储、传输、可视化、
and information privacy. /
和信息安全。/
The term often refers simply / to the use of predictive analytics or other certain advanced methods /
这个术语通常仅指 / 使用预测分析或者其他特定的高级方法 /
to extract value from data, and seldom to a particular size of data set [2]. /
从数据和少见的特殊大小的数据集合中提取数据。/

Internet of Things / 物联网

The Internet of Things (IoT) is the network of physical objects or “things” /
物联网是物体或“事物”组成的网络 /
embedded with electronics, software, sensors and connectivity / to enable it /
物体内置电子设备、软件、传感器和连通性 / 以激活它 /
to achieve greater value and service / by exchanging data /
获取更多价值和服务 / 通过交换数据 /
with the manufacturer, operator and/or other connected devices. /
与制造商、运营商和其他连接设备。/
Each thing is uniquely identifiable / through its embedded computing system /
每个物体都是唯一识别的 / 通过内置的计算系统 /
but is able to interoperate / within the existing Internet infrastructure [3]. /
但可以联动 / 在既存的网络基础设施内。/

Cloud Computing / 云计算

Cloud computing is a computing term or metaphor / that evolved / in the late 2000s, /
云计算是计算术语或者隐喻 / 它发展 / 在 2005~2010 年 /
based on utility and consumption of computing resources. /
基于计算资源的使用和消耗。/
Cloud computing involves deploying groups of remote servers and software networks /
云计算包括部署远程服务器和软件网络 /
that allow centralized data storage and online access / to computer services or resources. /
允许集中式数据存储和在线访问 / 计算机服务和资源。/
Clouds can be classified / as public, private or hybrid [1]. /
云可被划分为 / 公共云、私有云、混合云。/

技术段子摘选

Funny Technology Quotes

Cars will soon have the Internet on the dashboard. I worry that this will distract me from my texting.

— **Andy Borowitz**

"User" is the word used by the computer professional when they mean "idiot."

— **Dave Barry**

What did people do when they went to the bathroom before smart phones?

— **Aaron Cobra Mervis**

Thanks to the Internet, people we might have only suspected of being idiots can now give us ample evidence.

— **Andy Borowitz**

Technology is ruled by two types of people: those who manage what they do not understand, and those who understand what they do not manage.

— **Mike Trout**

Computers have enabled people to make more mistakes faster than almost any invention in history, with the possible exception of tequila and hand guns.

— **Mitch Ratcliffe**

Treat your password like your toothbrush. Don't let anybody else use it, and get a new one every six months.

— **Clifford Stoll**

I changed my password to "incorrect" so whenever I forget what it is, the computer will say "your password is incorrect."

— **Unknown**

Facebook is the only place on the planet where it is acceptable to talk to a wall.

— **Unknown**

I heard internet addiction's now an official mental disorder and you can go to rehab for it. I'm only going if there's Wi-Fi.

— **Unknown**

Modern technology. Owes ecology. An apology.

— **Alan M. Eddison**

There are two ways of constructing a software design; one way is to make it so simple that there are no deficiencies, and the other way is to make it so complicated that there are no obvious deficiencies. The first method is far more difficult.

— **C.A.R. Hoare**

Programming today is a race between software engineers striving to build bigger and better idiot-proof programs, and the Universe trying to produce bigger and better idiots. So far, the Universe is winning.

— **Rick Cook**

As IT professionals, we don't solve your problems. We just change the nature of problems.

— **Unknown**

I invented it, Bill made it famous.

— **David Bradley** (Wrote the code for Ctrl+Alt+Del on IBM PC)

My software never has bugs. It just develops random features.

— **Unknown**

I used to live an active life—I played football, tennis, did car racing. Sometimes played poker and pool. But when my computer got stolen, everything ended.

— **Unknown**

Unix is user-friendly. It's just very selective about who its friends are.

— **Unknown**

单词&短语

dashboard 仪表盘
idiot 傻子
mean 意味着
distract ~ from 使 ~ 分心
texting 收发短信
thanks to ~ 多亏 ~
suspect 怀疑
ample 充分的
evidence 证据
invention 发明
toothbrush 牙刷
get a new one 更换新的
possible exception 可能的例外
tequila 龙舌兰酒
hand gun 手枪
on the planet 在这个星球上
wall 墙壁
addiction 沉溺于
mental disorder 精神病
rehab 戒断治疗
ecology 生态学
apology 道歉
be impressed with 为 ~ 所感动
construct 构成
deficiency 缺乏
complicated 复杂的
obvious 明显的
method 方法
race 赛跑
idiot-proof 容易操作的
striving 努力
nature 本性
invent 发明
famous 著名的
random 随机
feature 功能
user-friendly 容易使用的
selective 选择的
active life 积极的生活
pool 台球
get stolen 丢失，被盗

根据提示完成句子

There are ___ ways / of constructing a ________ ______;
有两种方式/ 构成软件设计；/

one way is to ____ it so ______ /
一种是让其更简单/

that there are no ___________, /

无缺陷，/

and ___ _____ ___ is to ____ it so ___________ /

另一种方式是让其复杂/

that there are no _______ ____________. /

而无明显的缺陷。/

The first ______ is ___ more _________. /

第一种方法相对较难。/

— C.A.R. Hoare

Programming today is a ____ / between /

当今的编程是一场赛跑/ 间于/

software engineers ________ __ build /

软件工程师努力构建的/

bigger and better ___________ programs, /

更大、更易操作的程序，/

and the Universe ______ __ produce /

和宇宙尽力生产的/

bigger and better ______. /

更大、更好的傻子。

So far, / the Universe is winning. /

至今，/ 宇宙一直在获胜。

— Rick Cook

翻译

Funny Technology Quotes

技术段子摘选

Cars will soon have the Internet / on the dashboard. / I worry / that this will distract me /
汽车很快会有网络 / 在仪表盘上。/ 我担心 / 这会让我分心 /
from my texting. /
在收发短信时。/

——安迪・博罗维茨

"User" is the word / used by the computer professional / when they mean "idiot." /
"用户"是一个单词 / 被计算机专家使用 / 当他们表达"傻子"时。/

——戴维・巴瑞

What did people do / when they went to the bathroom / before smart phones? /
人们会做什么呢? / 当他们去卫生间的时候 / 在有手机之前 /

——艾伦・克布拉・摩尔维斯

Thanks to the Internet, / people we might have only suspected of being idiots /
多亏网络, / 我们曾经怀疑是"傻子"的那些人 /
can now give us ample evidence. /
现在给出了充分的证据。/

——安迪・博罗维茨

Technology is ruled / by two types of people: / those who manage /
技术被支配 / 由两种人: / 那些管理着 /
what they do not understand, / and those who understand / what they do not manage. /
他们不理解的东西, / 和那些理解着 / 他们不能管理的东西。/

——迈克・特劳特

Computers have enabled people / to make more mistakes faster /
计算机已经让人们 / 更迅速地犯更多错误 /
than almost any invention in history, / with the possible exception / of tequila and hand guns. /
比历史上的任何发明 / 此外是 / 龙舌兰酒和手枪。/

——米奇-雷克利夫

Treat your password / like your toothbrush. / Don't let anybody else use it, /
对待你的密码 / 就像你的牙刷。/ 不要让任何人使用, /
and get a new one / every six months. /
并且更新密码 / 每 6 个月。/

——克利福德・斯托尔

I changed my password / to "incorrect" / so whenever I forget / what it is, /
我将密码更新 / 为"错误" / 所以每当我忘记 / 密码时, /
the computer will say / "your password is incorrect." /
计算机会提示 / "你的密码错误"。/

——佚名

Facebook is the only place / on the planet / where it is acceptable / to talk to a wall. /
Facebook 是唯一的地方 / 在这个星球上 / 可以被接受的 / 对墙说话。/

——佚名

I heard internet addiction's now an official mental disorder / and you can go to rehab for it. /
我听说网瘾现在是公认的精神病 / 并且你可以去治疗。/
Modern technology. / Owes ecology. / An apology. /
现代技术。/ 欠生态学。/ 一个道歉。/

——艾伦·M. 爱迪生

There are two ways / of constructing a software design; / one way is to make it so simple /
有两种 / 构成软件设计的方式; / 一种是让其更简单 /
that there are no deficiencies, / and the other way is to make it so complicated /
无缺陷，/ 另一种方式是让其复杂 /
that there are no obvious deficiencies. / The first method is far more difficult. /
而无明显的缺陷。/ 第一种方法相对较难。/

——查尔斯·安东尼·理查德·霍尔爵士

Programming today is a race / between /
当今的编程是一场赛跑 / 间于 /
software engineers striving to build bigger and better idiot-proof programs, /
软件工程师努力构建的更大、更易操作的程序，/
and the Universe trying to produce bigger and better idiots. /
和宇宙尽力生产的更大、更好的傻子。/
So far, / the Universe is winning. /
至今，/ 宇宙一直在取胜。/

——里克·库克

As IT professionals, / we don't solve your problems. / We just change the nature of problems. /
作为 IT 专家，/ 我们不能解决你的问题。/ 我们只是改变问题的本质。/

——佚名

I invented it, / Bill made it famous. /
我发明了它，/比尔让它出名/. /

——大卫·布莱德利（在IBM个人电脑上编写了Ctrl+Alt+Del代码）

My software never has bugs. / It just develops random features. /
我的软件从来没有 Bug。/ 它仅是开发了随机功能。/

——佚名

I used to live an active life / - I played football, tennis, did car racing. /
我曾拥有积极的生活 / 踢足球、打网球、赛车。/
Sometimes played poker and pool. / But when my computer got stolen, / everything ended. /
有时打扑克和台球。/ 但我的计算机被偷时，/ 一切都结束了。/

——佚名

Unix is user-friendly. / It's just very selective / about who its friends are. /
Unix 很容易使用。/ 只不过它非常有选择性 / 对谁是自己的朋友。/

——佚名

自宫眼中的网络安全

White House on Cybersecurity

Through its website, the White House describes various issues affecting the country and outlines plans for addressing them. This is what the White House says about Cybersecurity and Internet Policy [1].

Cybersecurity and Internet Policy

President Obama has pledged to preserve the free and open nature of the Internet to encourage innovation, protect consumer choice, and defend free speech. The Administration has created an Internet Policy Task Force to bring together industry, consumer groups, and policy experts to identify ways of ensuring that the Internet remains a reliable and trustworthy resource for consumers and businesses.

In July 2011, at the Organisation for Economic Co-operation and Development (OECD), the Obama Administration joined with representatives from business, civil society, and Internet technical communities from 34 countries to reaffirm the importance of Internet policy principles that have enabled the open Internet to flourish with innovation and human connections beyond our wildest expectations.

Americans deserve an Internet that is safe and secure, so they can shop, bank, communicate, and learn online without fear their accounts will be hacked or their identity stolen. President Obama has declared that the "cyber threat is one of the most serious economic and national security challenges we face as a nation" and that "America's economic prosperity in the 21st century will depend on cybersecurity." To help the country meet this challenge and to ensure the Internet can continue as an engine of growth and prosperity, the Administration is implementing the National Strategy for Trusted Identities in Cyberspace. The Administration also released the International Strategy for Cyberspace to promote the free flow of information, the security and privacy of data, and the integrity of the interconnected networks, which are all essential to American and global economic prosperity and security.

President Obama has responded to Congress' call for input on the cyber-security legislation that our Nation needs, and the Administration will continue to engage with Congress as it moves forward.

The Obama Administration has also prioritized the cybersecurity of federal departments and agencies. In addition, the Administration has matured the government's implementation of the Federal Information Security Management Act (FISMA) away from a static, paper-based process to a dynamic, relevant process based upon continuous monitoring and risk assessment.

> "We have to do everything we can to encourage the entrepreneurial spirit, wherever we find it. We should be helping American companies compete and sell their products all over the world. We should be making it easier and faster to turn new ideas into new jobs and new businesses. And we should knock down any barriers that stand in the way. Because if we're going to create jobs now and in the future, we're going to have to out-build and out-educate and out-innovate every other country on Earth."
>
> -PRESIDENT BARACK OBAMA, SEPTEMBER 16, 2011

出 处

1. Technology, The White House, https://goo.gl/3Pcbv3

单词&短语

describe 描述
affect 影响
outline 提纲
address 从事，忙于
policy 政策
pledge to V 承诺 V
preserve 保存
nature 本性，本质
encourage 鼓励
innovation 革新
protect 保护
consumer 消费者
choice 选择
defend 防御
free speech 言论自由
expert 专家
identify 发现，识别
ensure 保证
remain 保持 ~
reliable 可信的
trustworthy 可信赖的
join with 与……结合
representatives 代表
civil 市民的，民间的
reaffirm 重申
principle 原理，原则
enable A to V 使 A 能够做 V
flourish 繁荣
expectation 期待
deserve 值得
bank 存款
account 账户
identity 身份
steal 偷
declare 宣言
serious 严重的
economic 经济的
national 国家的
challenge 挑战
face 直面
prosperity 繁荣
depend on 依据
continue 继续
growth 增长
implement 实现
release 发布
international 国际的
promote 促进
essential 必需的
respond to ~ 回应 ~
legislation 立法
engage with ~ 与 ~ 建立关系
prioritize 划分优先顺序
federal 联邦的
in addition 此外
mature 使成熟
static 静态的
dynamic 动态的
relevant 相关的
Based upon ~ 基于 ~
continuous monitoring 持续监控
risk assessment 风险评估
encourage 鼓励
entrepreneurial spirit 企业家精神

compete 竞争
turn A into B 将 A 换为 B
knock down 摧毁
barrier 障碍
stand in the way 妨碍
out- 更多地 –

根据提示完成句子

Americans ________ an Internet / that is _____ and _______, /
美国人值得拥有网络/ 安全和有保障的/

so they can _____, _____, ____________, /
这样他们可以购物、处理银行业务、通信/

and ______ _______ / without fear /
和在线学习/ 不用担心/

their _________ will be _______ or their _________ _______. /
自己的账户会被入侵或者身份被盗。/

President Obama has _________ / that the "_____ ______ is /
奥巴马总统已经宣布/ "网络威胁/

one of the most serious _________ and _________ security ___________ /
是最严重的经济和国家安全挑战之一/

we _____ as a nation" /
我们需要从国家层面应对" /

and that "America's _________ ___________ / in the 21st century /
并且 "美国的经济繁荣/ 在21世纪/

will _______ __ cybersecurity."
将取决于网络安全"。/

翻译

White House on Cybersecurity

白宫眼中的网络安全

Through its website, / the White House describes various issues / affecting the country /
通过网站，/ 白宫阐述各种话题 / 影响国家的 /
and outlines plans / for addressing them. /
和计划提纲 / 为解决它们的。/
This is / what the White House says about Cybersecurity and Internet Policy [1]. /
以下是 / 白宫眼中的网络安全和网络政策。/

Cybersecurity and Internet Policy / 网络安全和网络政策

President Obama has pledged to preserve the free and open nature of the Internet /
奥巴马总统发誓要保持网络自由和开放的本质 /
to encourage innovation, protect consumer choice, and defend free speech. /
以鼓励创新、保护消费者选择、捍卫言论自由。/
The Administration has created an Internet Policy Task Force /
政府已经成立一个网络政策专责小组 /
to bring together industry, consumer groups, and policy experts /
以聚集业界、消费者团体、政策专家 /
to identify ways of ensuring / that the Internet remains a reliable and trustworthy resource /
找出确保 / 网络仍然是一个值得信赖和依靠的资源 /
for consumers and businesses. /
对消费者和企业。/

In July 2011, / at the Organisation for Economic Co-operation and Development (OECD), /
2011 年 7 月，/ 在经济合作与发展组织会议上 /
the Obama Administration joined with representatives /
奥巴马政府和代表们合作
from business, civil society, and Internet technical communities /
来自商业、市民社会和网络技术共同体 /
from 34 countries / to reaffirm the importance of Internet policy principles /
34 个国家的 / 以重申网络政策原则的重要性 /
that have enabled the open Internet / to flourish with innovation and human connections /
这已经让开放的网络 / 振兴了创新和人类联系 /
beyond our wildest expectations. /
超出了我们的预想。/
Americans deserve an Internet / that is safe and secure, / so they can shop, bank, communicate, /
美国人值得拥有网络 / 是安全和有保障的 / 为此他们可以购物、处理银行业务、通信 /
and learn online / without fear / their accounts will be hacked or their identity stolen. /
和在线学习 / 不用担心 / 自己的账户会被入侵或者身份被盗。/
President Obama has declared /
奥巴马总统已经宣布 /

that the “cyber threat is / one of the most serious economic and national security challenges /
“网络威胁是最严重的经济和国家安全挑战之一 /
we face as a nation” / and that “America’s economic prosperity / in the 21st century /
我们要从国家层面应对” / 并且 “美国的经济繁荣 / 在 21 世纪的 /
will depend on cybersecurity.” / To help the country meet this challenge / and to ensure /
将取决于网络安全”。/ 为帮助国家应对挑战 / 和确保 /
the Internet can continue / as an engine of growth and prosperity, /
网络可以持续 / 成为发展和繁荣的动力，/
the Administration is implementing the National Strategy / for Trusted Identities / in Cyberspace. /
政府正在实施国家战略 / 为可信赖的人员 / 在网络空间。/
The Administration also released the International Strategy for Cyberspace /
政府也为网络空间发布了国际战略 /
to promote the free flow of information, the security and privacy of data, /
以促进信息的自由流通、数据的安全和隐私 /
and the integrity of the interconnected networks, /
以及联结网络的完整性，/
which are all essential / to American and global economic prosperity and security. /
所有一切都是必需的 / 对美国人和全球经济的繁荣和安全而言。/

President Obama has responded to Congress’ call / for input on the cybersecurity legislation /
奥巴马总统已经对国会的要求做出回应 / 对于提交的网络安全立法案 /
that our Nation needs, / and the Administration will continue to engage with Congress / as it moves forward.
我们国家需要的 / 并且政府将持续与议会接洽 / 随着日后的进展。/

The Obama Administration has also prioritized the cybersecurity of federal departments and agencies. /
奥巴马政府也划分了联邦部门和机关的网络安全优先顺序。/
In addition, / the Administration has matured the government’s implementation
of the Federal Information Security Management Act (FISMA) /
此外，/ 政府已经推动《联邦信息安全管理法案》政府实施走向成熟 /
away from a static, paper-based process / to a dynamic, relevant process /
从静态的书面过程 / 到动态的切实的过程 /
based upon continuous monitoring and risk assessment. /
基于持续性监控和风险评估。/

> “We have to do everything / we can / to encourage the entrepreneurial spirit, /
> “我们需要竭尽全力 / 我们能做的 / 以鼓励企业家精神，/
> wherever we find it. / We should be helping American companies compete and sell their products /
> 在我们能发现的所有地方。/我们应该帮助美国的公司竞争并出售他们的产品/
> all over the world. / We should be making it easier and faster / to turn new ideas /
> 在全球。/我们应该让其更容易、更快地/将新创意/
> into new jobs and new businesses. / And we should knock down any barriers /
> 转换为新职位和新业务。/我们应该击倒所有障碍/
> that stand in the way. / Because if we’re going to create jobs / now and in the future, /
> 挡在途中的。/因为如果我们计划创造职位/在现在和将来，/
> we’re going to have to out-build and out-educate and out-innovate every other country / on Earth.” /
> 我们需要比其他国家更有创造力、投入更多教育，不断创新/在地球上的。”/
>
> - PRESIDENT BARACK OBAMA, SEPTEMBER 16, 2011 /
> ——美国前总统贝拉克·奥巴马，2011年9月16日/

语法目录

· 提示：数字为单元序号，并非页码。

难易度目录

LEVEL 1 ★☆☆☆

LEVEL 2 ★★☆☆

LEVEL 3 ★★★☆

LEVEL 4 ★★★★